T0135732

Bibliografische Information der Deutschen Nationalbibliothek

Die Deutsche Nationalbibliothek verzeichnet diese Publikation in der Deutschen Nationalbibliografie; detaillierte bibliografische Daten sind im Internet über http://dnb.d-nb.de abrufbar.

ISBN 978-3-8325-2563-7

Logos Verlag Berlin GmbH
Comeniushof, Gubener Str. 47,
10243 Berlin
Tel.: +49 (0)30 42 85 10 90
Fax: +49 (0)30 42 85 10 92
INTERNET: http://www.logos-verlag.de

Logical Foundations of Database Transformations for Complex-Value Databases

Dissertation

zur Erlangung des akademischen Grades
Doktor der Naturwissenschaften
(Dr. rer. nat.)

der Technischen Fakultät
der Christian-Albrechts-Universität zu Kiel

Qing Wang

Kiel

2010

1. Gutachter	Prof. Dr. Bernhard Thalheim
2. Gutachter	Prof. Dr. Egon Börger
3. Gutachter	Prof. Dr. Klaus-Dieter Schewe
Datum der mündlichen Prüfung	28.05.2010

Abstract

Database transformations consist of queries and updates which are two fundamental types of computations in any databases - the first provides the capability to retrieve data and the second is used to maintain databases in light of ever-changing application domains. In the theoretical studies of database transformations, considerable effort has been directed towards exploiting the close ties between database queries and mathematical logics. It is widely acknowledged that a logic-based perspective for database queries can provide a yardstick for measuring the expressiveness and complexity of query languages. Furthermore, mathematical logics encourage the expedited development of declarative query languages that have the advantage of separating the logical concerns of a query from its implementation details. However, in sharp contrast to elegant theories of database queries resulting from extensive studies over the years, the understanding of logical foundations of database updates is paltry.

With the rising popularity of web-based applications and service-oriented architectures, the development of database theories in these new contexts must address new challenges, which frequently call for establishing a theoretical framework that unifies both queries and updates over complex-value databases. More specifically, in rich Web application architectures, queries themselves are not sufficient to support data processing; interactive integration among Web-accessible services requires the compositionality of queries and updates; the increasing complexity of application domains demands more flexible data structures than ubiquitous relations, which leads to complex-values represented by arbitrary nesting of various type constructors (e.g., set, list, multiset and tuple). Therefore, a theoretical framework of database transformations plays an important role in investigating a broad range of problems arising from extensions of query languages with update facilities, such as, database compilers and optimisers.

To date, there has been only limited research into a unifying formalisation of database queries and updates. The previous findings reveal that it is very difficult to characterise common features of database queries and updates in a way which is meaningful for further theoretical investigations. However, the advent of the sequential Abstract State Machine (ASM) thesis capturing sequential algorithms sheds light on the study of database transformations. Observing that the class of computations described by database transformations may be formalised as algorithms respecting database principles, I am inspired by using abstract state machines to characterise database transformations. In doing so, this dissertation aims to lay down the foundations for establishing a theoretical framework of database transformations in the context of complex-value databases.

My first major contribution in this dissertation is to propose a complete characterisation of database transformations over complex-value databases from an algorithmic point of view. Five intuitive postulates are defined for highlighting the essence of database transformations. Furthermore, a formal computation model for database transformations, called database Abstract State Machines (DB-ASMs), is developed. It turns out that every database transformation characterised by the postulates can be behaviourally simulated by a DB-ASM with the same signature and background, and vice versa. My second major contribution is a logical formalisation for DB-ASMs. In spite of bounded non-determinism permitted by DB-ASMs, the logic for DB-ASMs is proven to be sound and complete. This is due to the finiteness condition stipulated on the database part of a state, which thereby leads to the finiteness of update sets in one-step transitions. These findings empower the use of various verification tools for studying the properties of database transformations over complex-value databases.

To identify different subclasses of database transformations, I investigate the customisation of backgrounds. The relational and tree-based backgrounds are formalised for characterising relational and XML database transformations, respectively. For the relational backgrounds, I focus on the connection between the constructivity of backgrounds and the representation of relational algebra which is widely used for query rewriting and optimisation at an internal implementation level in commercial database systems. Furthermore, I develop an elegant computation model for XML database transformations, called XML machines, which incorporates weak Monadic Second-Order (MSO) logic into DB-ASM rules and can specify XML database transformations at a flexible abstraction level. It is found that incorporating MSO logic into DB-ASM rules can not actually increase the expressiveness of XML machines.

Finally, I address the partial update problem in the context of complex-value databases. In database transformations over complex-value databases, bounded parallelism is intrinsic and complex data structures form the core of each data model. Thus, the problem of partial updates arises naturally. Due to the ability to arbitrarily nest type constructors in a data model, the assumption on the disjointness of locations must be lifted. While, in principle, locations defined in a standard way bound to complex values are independent from each other, I also consider each position within a complex value as a location. This extension naturally leads to a dependency relation among locations. Then, I propose an efficient approach for checking the consistency of a given set of partial updates, which involves two stages. The first stage uses an algebraic approach to normalise shared updates based on the compatibility of operators, while the second stage checks the compatibility of clusters by integrating exclusive updates level-by-level. I show that partial updates can be applied in aggregate computing for optimisation.

Acknowledgement

I would like to express my gratitude to many people who were in direct or indirect contact with my work on this dissertation.

Most of all I owe my deepest gratitude to Klaus-Dieter Schewe, my supervisor, who has been a continuous source of inspiration. He opened up my eyes to the academic world. He gave me the freedom to pursue my research interests; in the meantime, he has always been there to help me through all the tough challenges I had met. He introduced me to the theory of abstract state machines and brought me to the field of logic and mathematics. He taught me how to use scientific methods to solve a problem. Without his patient guidance and stimulating suggestions, this dissertation would not have been successful completed. My gratitude for his trust, patience and generosity goes beyond words.

I am also very grateful to Bernhard Thalheim for all kinds of support he has provided during the past four years. I have been fortunate to have the opportunity to work with him together, and have learned a great deal from his extensive knowledge, his thoughtful recommendations and his valuable comments on my work.

I would like to show my gratitude to Jose Maria Turull Torres, my co-supervisor, for his interest in my work. His feedback helped me to make improvements to the dissertation, especially, at the early stage of my PhD studies.

I would like to thank Sven Hartmann. He was always available for support and advices. He was of a great help in difficult times. A very special thanks and mention to Thu Trinh for her warmhearted friendship. I will never forget good times we had together at the beautiful Lipari Island. She kindly helped me proofread and sharpen up the final version of the dissertation. A special thanks also goes to Tracy Norrish for helping correct grammatical errors. I wish to thank many my former colleagues at Information Science Research Centre (ISRC) who provided me with the friendly assistance in a number of ways.

My work was partially funded by the Tertiary Education Commission of New Zealand. I would like to express my sincere thanks here. The funding not only freed me from the financial worries for three years but also paid for several research trips that helped enhance my research experience.

I also would like to acknowledge the support from my employer, the University of Otago, for granting me the leave to complete my PhD studies during the final months.

Lastly, and most importantly, I am forever indebted to my family for their love, understanding and endless support when it was most required. I dedicate my dissertation to you.

Contents

Chapter 1

Introduction

This chapter serves three purposes: motivating the research problems, defining the research objectives and summarising the significant research outcomes.

First of all, I discuss the motivation for considering database queries and updates in a unifying framework. The interactive and intimate relationships between queries and updates in several database paradigms are demonstrated via a running example. Several difficult issues in establishing such a unifying framework are highlighted. Then I present the major objectives of this dissertation. In particular, I clarify the scope of the research by identifying the class of computations referred to by the term *database transformation* and discuss several reasons why Abstract State Machine is appropriate as a methodology for investigating the characterisation of database transformations. Finally, a detailed overview for subsequent chapters is provided. The focus is particularly on the problems encountered during my investigation, the solutions proposed for resolving the problems and the summaries of the main results.

1.1 Motivation

The study of database queries has always been a central theme for database theoreticians. From a computational point of view, a query to a relational database transforms database instances over an input schema into database instances over an output schema, in which input and output schemata are considered to be completely independent from each other. Since the introduction of relational calculus – a fragment of the first-order logic – as a query language over relational databases, investigations on the logical grounds of database queries have attracted much attention from database communities. In theoretical studies, it has been well acknowledged that a logic-based perspective for database queries can provide a yardstick for measuring the expressiveness and complexity of query languages. Additionally, an immediate consequence of applying mathematical logics in the database area is the expedited development of declarative query languages that has the advantage of separating the logical concerns of a query from its implementation details. When the logic of a query is described by certain declarative query language, the issues of query optimisation can be handled by a database system with little to rely on the expertise of database developers. More specifically, a database system first translates a declarative query into an algebraic expression, then applies various optimisation rules predefined in the system to rewrite an algebraic expression, and finally executes the optimised algebraic expression with improved database performance. In relational database systems, relational algebra is such an algebraic language facilitating the optimisation of relational database queries.

However, query languages themselves are not sufficient for most database applications. Because the world modelled by databases is not static, database instances themselves change over time. Thus, updates effectuating these changes play an important role, and often, database queries and updates turn out to have intimate connections. For instance, a relation may be updated by using matched tuples from another (possibly the same) relation on a join operation over several specified attributes, tuples of a relation may be deleted based on some selection criteria, etc. On the one hand, such connections between queries and updates justify the support for the embedding of queries in updates as a fundamental feature in all major commercial relational database systems. On the other hand, it brings up considerable concerns on the theoretical foundations for database updates. In sharp contrast to the elegant and fruitful theories for database queries, the theoretical foundations for database updates, and more generally for a unifying framework encompassing both queries and updates are still lacking.

The following example is taken from an Oracle database application. It illustrates the close relationship between queries and updates in relational databases.

Example 1.1.1. Consider a relational database schema consisting of the following four relation schemata: PUBLICATION, PERSON, ACADEMICUNIT and AUTHORSHIP such that

- PUBLICATION = {PubID,Title,Year,Category,File,Mime,NoOfDownload},
- PERSON = {PersonID,Fname,Sname,Address},

- AcademicUnit = {UnitID,Name,ParentID,OldID} and
- Authorship = {PubID,UnitID,PersonID,Order}.

Assume that the authorship of publications is still associated with old academic unit IDs stored in the attribute OldID of AcademicUnit. To change all the academic unit IDs kept in the attribute UnitID of AcademicUnit to the new unit IDs, we may use the following statement[1].

```
UPDATE Authorship a
SET a.UnitID = ( SELECT b.UnitID
                 FROM AcademicUnit b
                 WHERE a.UnitID = b.OldID);
```

□

In the past decades, the interests arising from the development of object-oriented applications and programming languages that utilise objects have driven the studies of object-based databases. There are several principal arguments for justifying an object-based approach to database systems: (i) to enrich the ability of modelling various business objects in a database; (ii) to minimise the possible mismatch between the data model used in an application and the data model supported by a database; (iii) to support the native management of objects in a database. The emergence of query languages with extended technologies for handling object-based databases [3] required the view on database queries to be somehow relaxed: a query is considered as a binary relation from database instances over an input schema to database instances over an extended output schema which preserve the database instance over the input schema. This relaxation reflects the significance of object creation, during a computation and in a final result, for increasing the expressiveness of query languages and the capability of data modelling. Behind the scenes, object identifiers play an essential role in object-based databases, functioning as a powerful programming and modelling primitive for set manipulation, structure sharing, update or the encoding of cyclicity [3]. An object identifier is assigned to each object at the time of creation. As object identifier is an implementation notion for internal complex structures [3, 124], it leads to the non-deterministic choice on object identifers. Therefore, the need of assigning object identifiers to new objects blurs the distinction between queries and updates. More specifically, a database computation accomplishing certain functionality in the presence of

[1] Assume that there is a unique constraint on the attribute OldID of AcademicUnit.

objects may be formalised as a query in one language, as an update in another language, or as the mix of queries and updates in a language different from both of them.

The following example illustrates how a computation task can be implemented as queries and updates in different database languages, respectively.

Example 1.1.2. Let us consider the relation schema AcademicUnit provided in Example 1.1.1 again. Suppose we have an inclusion constraint stating that values in the attribute ParentID should be values in the attribute UnitID. Then AcademicUnit may contain recursive structures. Instead of modelling academic units as tuples in a relation over AcademicUnit = {UnitID,Name,ParentID,OldId}, they may also be modelled as objects in a class AcademicUnits such that each object has a unique object identifier (i.e., a unit id) together with an unit name, a set of its child units and an old unit id.

Let S_1 and S_2 be two database instances representing the same set of academic units from a relational point of view and from an object-based point of view, respectively. Transforming the database instance S_1 into the database instance S_2 is a common task undertaken by database applications. We will show that this task may be implemented in several different ways.

Let us first have a look at how it can be achieved in Oracle applications. Oracle provides the object technology which is a thin layer of abstraction built over the relational technology.

```
CREATE TYPE tAcademicUnit;

CREATE TYPE tAcademicUnits AS table of ref tAcademicUnit;

CREATE TYPE tAcademicUnit AS object (
                Name     varchar2,
                ChildID tAcademicUnits
                OldID     varchar2 );

CREATE TABLE AcademicUnits (
                AcademicUnit tAcademicUnit );

INSERT INTO AcademicUnits
      SELECT tAcademicUnit(Name, null, OldID)
        FROM AcademicUnit;
```

```
UPDATE AcademicUnits b
    SET b.AcademicUnit.ChildID=(
        CAST(
            MULTISET(
                SELECT ref(a)
                    FROM AcademicUnits a,
                         AcademicUnit c,
                         AcademicUnit d
                WHERE a.AcademicUnit.Name = c.Name
                    AND d.UnitID = c.ParentID
                    AND d.Name = b.AcademicUnit.Name)
        AS tAcademicUnits)
);
```

To accomplish such a transformation, a more powerful query language *Identity Query Language* (IQL) [3] can also be used. As demonstrated by Example 1.2 in [3], there exists a desired query for this transformation in which four stages occur sequentially or in parallel with inflationary semantics [1, 3]:

(1) A set of distinct names of academic units is produced;

(2) A pair of object identifiers per name is produced such that the first one is assigned to an object representing an academic unit while the second one is assigned to an object which will have a set of child academic unit objects but is still empty at this stage;

(3) For two objects corresponding to a pair of object identifiers, object identifiers of objects representing child academic units of the first object are nested into the value of the second object;

(4) A set of objects representing academic units are generated as desired, whose values consist of name of an academic unit, a set of object identifiers referring to its child academic unit objects and an old unit id. □

In recent years, web-based applications are rapidly emerging. A lot of research focuses on investigating service-oriented architectures (SOA), treating functionality as interoperable services, in particular, web services. Broadly speaking, web services can in fact be anything relating to software systems: simple functions, database manipulations, Application Programming Interfaces or fully functional Web Information Systems. By means of exposing a database and the data logic in a database as web services, well-known database vendors such as Oracle, IBM, and Microsoft, etc. provide database web services technology as a data-centric approach to web services. In general, there are two viewpoints for database web services: as service provider to allow the access to a database via web services mechanisms and as service consumer to allow a database program such as a database query or update to invoke a web service. A database can simultaneously play the role of a service provider and consumer. With the desired characteristic that services can be easily assembled into more complex (possibly unanticipated) services, database web services should be closed under compositions. However, in order to achieve such compositionality among queries and updates, a formal framework for database computations unifying both queries and updates needs to be established.

The following scenario demonstrates the close interplay between queries and updates during a composition of web services implemented as Oracle database applications.

Example 1.1.3. Based on the relational database schema presented in Example 1.1.1, let us consider two web services DOWNLOADPUBLICATION and FINDPUBLICATION:

- DOWNLOADPUBLICATION first downloads the content of publications, and then increases values in the attribute NoOfDownload of PUBLICATION for these publications by 1;
- FINDPUBLICATION finds the ids of publications according to a condition on the attribute Title of PUBLICATION[2].

The following is one possible implementation of the two services using Oracle PL/SQL, where Oracle provides a mechanism WPG_DOCLOAD to download content files directly from a database.

```
TYPE PubidType IS ref cursor RETURN Publication.PubID%TYPE;

PROCEDURE downloadPublication(
    IDcursor IN PubidType)
AS
```

[2]This service can be generalised to be a search engine service based on more complicated conditions. To simplify the discussion, here only the condition on the attribute Title is considered.

```
        vPubID PUBLICATION.PubID%TYPE;
        vFile PUBLICATION.File%TYPE;
        vMime PUBLICATION.Mime%TYPE
    BEGIN
        OPEN IDcursor;
        LOOP
            FETCH IDcursor
              INTO vPubID;
              EXIT WHEN IDcursor%NOTFOUND;
              SELECT File, Mime
                 INTO vFile, vMime
                FROM PUBLICATION
              WHERE PubID = vPubID;

            OWA_UTIL.mime_header(vMime,FALSE);
            HTP.p('Content-length: ' || DBMS_LOB.getlength(vFile));
            OWA_UTIL.http_header_close;

            WPG_DOCLOAD.download_file(vFile);

            UPDATE PUBLICATION
                  SET NoOfDownload=vNoOfDownload+1
             WHERE PubID=vPubID;
            COMMIT;
        END LOOP;
        CLOSE IDcursor;
    END DOWNLOADPUBLICATION;

    PROCEDURE FINDPUBLICATION(
        Description IN VARCHAR2,
        IDcursor OUT PUBIDTYPE)
    AS
    BEGIN
        OPEN IDcursor FOR
              SELECT PubID
                FROM PUBLICATION
```

WHERE Title like '%' || Description || '%';
END FINDPUBLICATION;

A web service GETPUBLICATION that downloads the content of publications whose titles match some entered description can be created as a composition of two web services DOWNLOADPUBLICATION and FINDPUBLICATION as follows:

DOWNLOADPUBLICATION(FINDPUBLICATION(Description)).

In order to show the details of publications after downloading them, the above web service GETPUBLICATION may be further composed with another web service called SHOW-PUBLICATION which may present the metadata of these publications along with the up-to-date numbers of downloads. □

In database theory, the term *database transformation* is used to refer to a unifying treatment for computable queries and updates. More precisely, a database transformation is defined as a binary relation on database instances over an input schema and database instances over an output schema in which certain criteria should be satisfied, such as well-typedness, effective computability, genericity and functionality [2, 10]. Unifying queries and updates into a general framework is very appealing to database communities, however, it remains a challenging problem. A great amount of research aimed at formalising different classes of database transformations, e.g. [3, 146, 148, 150]. Nevertheless, very little progress has been made with respect to understanding the theoretical foundation of database transformations. Many investigations have yielded meaningful findings in the case of queries. Unfortunately, extending these results to updates is by no means straightforward [149]. In this dissertation, I investigate database transformations from an algorithmic point of view. Hence, in the remainder of the dissertation, the term "database transformation" is in fact used to refer to database transformation algorithms.

Abstract State Machines (ASMs) provide a universal computation model that formalises the notion of (sequential or parallel) algorithm [27, 82]. In Gurevich's seminal work [82] on the sequential ASM thesis, he pointed out the differences between a computable function in the recursion-theoretic sense and an algorithm. Strictly speaking, many algorithms in numerical mathematics, e.g., Newton's algorithm for determining zeros of a differentiable function $f : \mathbb{R} \rightarrow \mathbb{R}$, do not define computable functions, as they deal with non-denumerable sets. Only when restricted to a countable subset – such as floating-point numbers instead of real numbers – and properly encoded can algorithms be considered as computable functions. Analogous to the fact that an algorithm is more than a computable function, there is more to a database transformation algorithm than a database transformation defined over input and output database instances.

In formalising a unifying framework for database transformations, several issues have to be taken into consideration. As raised in [149], the compositionality of database transformations is a difficult problem in the object-oriented paradigm. This is due to the unique identification of objects by object identifiers. The connection between an object and its object identifier is permanent, meaning that object identifiers can never be separated from objects once they have been assigned. However, the mechanism of assigning object identifiers to objects is entirely controlled by the underlying system, which makes objects difficult to be arbitrarily composed in the way that values can. The second issue is the non-determinism caused by the de facto view on object identifiers, i.e., the representation of object identifiers is irrelevant and only the interrelationship between objects matters [2]. In fact, the degree of non-determinism in a class of database transformations is one of critical factors which determine the upper bound of the expressiveness of associated languages. Apart from these two issues, it has also been perceived that there is a mismatch between the declarative semantics of query languages and the operational semantics of update languages. Unlike many query languages which can describe what a program should accomplish rather than specify how to accomplish, update languages usually require explicit specifications of operations and control flow in order to handle a problem. The obvious question is how to integrate different semantics within one database language?

To the best of my knowledge, there does not exist a formal computation model specifying both queries and updates over complex-value databases at a flexible abstraction level. Finding such a formal computation model is an important starting point for the formal investigation into theoretical properties of database transformations.

1.2 Objectives

The goal of this dissertation is to establish a formal and unifying framework for computations over complex-value databases. To accomplish this goal, the following objectives are set.

- The first and most important objective is to study how database transformations can be characterised in a robust mathematical way and to consider what kind of computation model is an appropriate formalism for database transformations.

- The second objective is to study the logical reasoning for properties of the computation model formalised in the first objective. In particular, I aim at establishing a sound and complete proof system for the logical characterisation proposed for database transformations.
- The third objective is to study two important subclasses of database transformations: relational and XML database transformations within the theoretical framework established in this dissertation.
- The last objective is to investigate the partial update problem arising from database transformations on complex-value databases.

My investigation is governed by the following: (i) what class of computations is meant by the notion of database transformation used in this dissertation; (ii) why the methodology of ASMs can serve as an appropriate foundation for the study of database transformations. I will discuss them in the following subsections.

1.2.1 Scope

To clarify the scope of this research, it is important to identify the class of computations called *database transformations*. Generally speaking, there are several principles that distinguish database transformations from other classes of computations in one or more aspects. Firstly, database transformations are abstract by nature, i.e., regardless of specific representations, database transformations are only concerned with structural properties of a database. This abstract point of view on computations is generally identified as the *genericity principle* in database communities. A variety of definitions of genericity have been proposed with the aim of classifying database computations into different subclasses. In this respect, I will take a more liberal view: preserving the genericity of database transformations only in terms of equivalence of substructures. Furthermore, database computations are typically considered over finite structures. This perspective is motivated by the finiteness of practical databases, and it then enriches theoretical knowledge about finite structures, for example, finite model theory, descriptive complexity theory, etc. Although finiteness is a dominant view for database structures, it fails to describe structural parts that are needed in connection with algorithmic properties. In order to study database transformations in a framework that has all the capability of dealing with algorithmic issues, I extend the view on a structure of databases from a finite structure to a meta-finite structure [74] consisting of a finite database part and an infinite algorithmic part interconnected in a rather restricted manner. Furthermore, to account for computations over all sorts of data models in this framework, database transformations in this dissertation are studied in a rich context of complex-value databases, which ideally may serve as the upper bound of other data models used in database applications.

In terms of various programming languages, database transformations represent a class of computations lying between general-purpose programming languages (such as C, C++, Jave, PhP, etc.) which are Turing-complete, and domain specific languages (such as SQL, RC, RA, XPath, etc.) which are designed for particular application domains. On one side, general-purpose programming languages do not have an ability to efficiently process information stored in a database; on the other side, domain specific languages may lack strong flow of control. A selective combination of functionalities from both sides while still respecting database principles is the key in characterising the class of computations represented by database transformations.

1.2.2 Methodology

Abstract State Machine (ASM) is a formal specification language. The sequential ASM thesis [82] shows that it can capture all the sequential algorithms, thus ASMs are accepted as a universal computation model formalising the notion of algorithm. In this study, the ASM methodology sheds light on how to establish a formal framework for database transformations. Intuitively, the idea originates from the observation that the class of computations described by database transformations may be formalised as a customisation of ASMs respecting database principles.

There are several good reasons why ASMs are particularly suitable for specifying database transformations.

One of the main advantages of ASMs is the flexibility of specification at any level of abstraction. It thus provides the capability of smoothly combining the declarative purity of queries with the procedural style of updates. More precisely, ASMs allow a mixed style in which declarative and procedural semantics can be used for different parts of a computation model according to specific requirements. Extending a query language by incorporating additional update functionality may lead to a more powerful language that has declarative parts spread over procedural parts, e.g., Data Manipulation Language in SQL, XQuery with update facility, etc.

ASMs also allow for a refinement of a high level specification for database transformations into a level where transactions can be made explicit [47, 101]. Therefore, a database transformation can be initiated from the end users' point of view by using ASMs to specify the ground requirements. After that, through iterative refinement procedures the specified database transformation can be fine-tuned into an executable database transaction which may consist of one or more update and query statements.

In terms of computability, ASM is a Turing-complete language. Therefore, ASMs are powerful enough for investigating problems that are difficult in the field of database research, as long as certain restrictions are appropriately imposed to ensure the compliance of database principles. Moreover, it is also possible to identify a fragment of ASMs satisfying some properties and then use its strengths for investigating more specific problems. The findings in the sequential and parallel ASM theses [27, 82] exemplify such an approach to precisely characterise different notions of algorithms. In this sense, by considering database transformations as special kinds of algorithms, ASMs shed light on how to characterise database transformations.

1.3 Detailed Overview

The following is a detailed description of what will be presented in the remainder of this dissertation.

Chapter 2 provides a literature review on the related works. In particular, a historical development of the theory on database transformations, several relevant characterisation theorems of ASMs, typical abstract computation models developed in database theory and their connections to logics are presented.

Chapter 3 addresses the question of how ASMs can be used to characterise database transformations in general. I start by examining database transformations in the light of the postulates for sequential algorithms[3] defined in [82]. As a result, five intuitive postulates for database transformations are defined: the sequential time postulate, the abstract state postulate, the background postulate, the bounded exploration postulate and the bounded non-determinism postulate.

Pragmatically, database transformations must terminate. This implies that a run will always be finite and a final state can always be reached. Furthermore, in order to take into

[3]In Gurevich's theory sequential algorithms permit only bounded parallelism, whereas parallel algorithms are understood to capture even unbounded parallelism.

consideration not only deterministic but also non-deterministic database transformations, the requirement of a one-step transition function has to be relaxed. For this, a one-step transition relation over states is permitted, leading to a slightly modified *sequential time postulate*. The necessity for non-determinism arises, among others, from the creation of objects [148]. Nevertheless, the degree of permissible non-determinism will be severely limited by further postulates.

According to the abstract state postulate for ASMs, states are first-order structures and the sets of states used for an algorithm are invariant under isomorphisms. In general, this notion of state must capture all databases, and in particular, we must take the following issue into account: not only logical structures in databases but also auxiliary structures for computing need to be captured in a state. Therefore, I will customise the *abstract state postulate* by requiring that a state is composed of a finite database part and a possibly infinite algorithmic part linked via bridge functions. This approach indeed engages a fundamental idea from meta-finite model theory [74].

The third postulate in the sequential ASM thesis, the bounded exploration postulate, requires that there is a finite set of terms (i.e., the bounded exploration witness) and only these can be updated in a one-step transformation. The generalisation of this postulate to the case of parallel algorithms in the parallel ASM thesis [27] leads to several significantly more complex postulates. As database transformations are intrinsically parallel computations, though an implementation may be sequential, I will adopt parts of these more complex postulates. The adoption will only be partial, as the parallelism in database transformations is rather limited in practice; it merely amounts to the same computation on different sets of data.

Analogous to the parallel ASM thesis, the background of a computation, which contains everything that is needed to perform the computation, but is not yet captured by the states, has to be explicitly stated. For instance, truth values and their connectives, and a value $\perp$ to denote undefinedness constitute necessary elements in a background. Furthermore, for database transformations, constructs that are determined by the used data model must be captured, so type constructors will have to be considered, along with functions defined on such types. This leads to the *background postulate* for database transformations.

Regarding the restricted form of parallelism needed for database transformations, I capture this by location operators, which generalise aggregation functions and cumulative updates. With these location operators I actually deal with meta-finite structures with multiset operations as defined in [74]. In doing so, update multisets have to be considered, which are then reduced to update sets by means of applying location operators. Furthermore, depending on the data model used and thus on the actual background signature, complex values (e.g., tree-structured values) may be used, which gives rise to the problem of partial updates [89]. That is, how to ensure that parallel updates to different parts of a complex-value (e.g., a tree) are synchronised. Dealing with partial updates actually is encompassed in the notion of consistent update set. Taking these ingredients together, a slightly modified *bounded exploration postulate* is obtained.

The fifth postulate, called the *genericity postulate*, addresses how the longstanding genericity principle from database theory limits the degree of non-determinism allowed in computations. My approach is to require all equivalent substructures to appear as substructures in one of the states reachable by a one-step transition whenever a substructure is preserved by the one-step transition. With the involvement of non-determinism, equivalence of substructures may indeed be destroyed, but the genericity postulate ensures that this can only be done in very limited ways, for example, by creating new identifiers.

In the rest of this chapter, a variant of Abstract State Machines, called *database Abstract State Machines* (DB-ASMs), is defined. Naturally, the permitted non-determinism requires the presence of a choice-construct, while the restricted parallelism leads to a let-construct that binds locations to location operators. I first show that DB-ASMs satisfy the five postulates for database transformations and proceed to show the converse that every database transformation stipulated by the postulates can be behaviourally simulated by a DB-ASM. This establishes the correctness of the DB-ASM formalisation.

Two important outcomes in this chapter are:

Theorem 3.2.1 Each DB-ASM M defines a database transformation T with the same signature and background as M.

Theorem 3.3.1 For every database transformation T, there exists an equivalent DB-ASM M.

In **Chapter 4** two specific theoretical frameworks for relational and XML database transformations are developed, built upon the general characterisation of database transformations by DB-ASMs. This leads to two major contributions: Firstly, the relational and tree-based backgrounds that are necessary for modeling relational and XML data are presented, which tailor DB-ASMs for capturing relational and XML database transformations, respectively. Secondly, a more elegant computational model, called XML machines, is developed which mainly differs from the DB-ASM model by the use of weak monadic second-order logic in forall and choice rules. It is shown that XML machines are, in fact, behaviourally equivalent to the DB-ASM model with tree-based backgrounds.

Theorem 4.5.1 XML Machines with a tree-based background that consists of a tree background class K^{tree} and a set $\tilde{\Gamma}^{tree}$ of tree type schemes capture exactly all XML database transformations with the same signature and background.

The specific backgrounds for relational and XML database transformations are formalised separately as follows:.

- For the relational backgrounds of relational database transformations, I begin with an introduction to the relational model, i.e., the Relational Data Model (RDM) and the Nested Relational Data Model (NRDM). After defining relational background classes and relational type schemes, the focus is on investigating the constructivity of relational backgrounds within the framework defined by Blass and Gurevich [29]. I exploit that the use of relational algebra in relational databases for query execution and optimisation at an internal implementation level requires its algebraic operators and set of algebra identities to be included in the relational backgrounds for relational database transformations.

- For the tree-based backgrounds of XML database transformations, I start by defining unranked trees using child and sibling relations. This is tailored towards XML trees by labelling and value functions. In order to provide more flexible operations on XML trees, contexts are introduced to support the freedom of selecting tree portions of interest, along with the use of subtrees. Furthermore, a many-sorted algebra on XML trees is defined, i.e. list of trees, contexts and labels, extending known tree algebras [41, 157]. Due to the finiteness of the database part in a state, XML trees are assumed to be always finite. Therefore, hereditarily finite trees with the functions required for trees and the functions used in hedge algebra operations together form the major components for tree-based backgrounds. Furthermore, in order to capture schemata for XML documents, I adopt extended document type definitions (EDTDs) [116], which, according to [111], generalise many other XML schema definition languages. These further add tree typing schemes to tree-based backgrounds. Nevertheless, for XML database transformations only initial and final states adhering to given typing schemes are required.

The second contribution in this chapter is the development of an alternative model of computation for XML database transformations, which exploits weak monadic second-order logic (MSO). Pragmatically speaking, the use of weak MSO formulae in forall and choice rules permits more flexible access to the database. Moreover, weak MSO logic is linked to regular tree languages [65, 140]: a set of trees is regular iff it is in weak MSO logic with k successors. As XML is intrinsically connected with regular languages, a lot of research has been done to link XML with automata and defining logics. It turns out that (weak) MSO logic has stood out as being a natural and important logic in such computations over tree-based structures, including XML database transformations.

As weak MSO logic subsumes first-order logic, it is straightforward to see that this natural model of XML machines captures all transformations that can be expressed by the DB-ASM model with tree-based backgrounds. As the result in Theorem 3.3.1 states that DB-ASMs capture all database transformations as defined by the five intuitive postulates, it should also not come as a surprise that XML machines are in fact equivalent to DB-ASMs with tree-based backgrounds. Since the hard part of the proof is already captured by the main characterisation theorem in Theorem 3.3.1, it remains only to show that XML machines satisfy the five postulates.

Chapter 5 presents a logical characterisation for DB-ASMs, and thus for database transformations in general, in accordance with the results in Chapter 3. The main contribution of this chapter is the establishment of a sound and complete proof system for the logic for DB-ASMs, which can be turned into a tool for reasoning about database transformations.

The following theorems have been proven in this chapter in relation to the proof system as defined in Section 5.3.

Theorem 5.4.1 (soundness) Let M be a DB-ASM and Φ be a set of sentences. If φ is derivable from a set Φ of formulae with respect to M (i.e., $\Phi \vdash_M \varphi$), then a formula φ is implied by a set Φ of formulae with respect to M (i.e., $\Phi \models_M \varphi$).

Theorem 5.5.1 (completeness) Let M be a DB-ASM and Φ be a set of sentences. If a formula φ is implied by a set Φ of formulae with respect to M (i.e., $\Phi \models_M \varphi$), then φ is derivable from a set Φ of formulae with respect to M (i.e., $\Phi \vdash_M \varphi$).

Since the rules of DB-ASMs exclude call rules, there is no iteration involved in each computation. Therefore, DB-ASMs may be treated in a similar way to hierarchical ASMs defined in [134] with the following extensions:

- states defined as meta-finite structures, instead of first-order structures,
- restricted non-determinism permitted by using only database variables in choice rules which range over the database part of a meta-finite state, and
- multiset operations provided by using ρ-terms and let rules over meta-finite states.

Consequently, the logic for DB-ASMs extends the logic for ASMs presented in [134] in several aspects:

- Firstly, the logic for DB-ASMs is built upon the logic of meta-finite structures, which is beyond the first-order logic due to the introduction of multiset operations. The formalisation of multiset operations is captured by the notion of ρ-terms that allows an inductive construction between formulae and terms. The use of ρ-terms greatly enhances the expressive power of the logic for DB-ASMs since aggregate computing in database applications can be easily expressed by using ρ-terms. Nevertheless, the presence of ρ-terms also greatly increases the complexity of the completeness proof for the proposed proof system. To handle this problem, I will use an inductive argument to construct structures based on the nested depth of ρ-terms. This piece of work is presented in Section 5.5.
- Secondly, the non-determinism accompanied with the use of choice rules poses a further challenging problem. As discussed in [134] for the logic for ASMs, it is improbable that there is a natural and simple approach to introduce non-determinism into the logic for ASMs. Fortunately, DB-ASMs are restricted to have qualifiers only over the database part of a state which is a finite structure. This implies that any update set or multiset yielded by a DB-ASM rule must be finite. Based on the finiteness of update sets, I use the modal operator $[\Delta]$ for each update set Δ generated by a DB-ASM rule r, rather than the modal operator $[r]$ for a rule r. By introducing $[\Delta]$ into the extended formulae of the logic for DB-ASMs, it is shown that non-deterministic database transformations can also be captured.

Chapter 6 examines the problem of partial updates in the context of complex-value databases. In many database applications over complex-value databases, the notion of location lends itself to a variety of perspectives based on the underlying data models. Therefore, to improve the naturalness and efficiency of database transformations in complex-value databases, I extend the notion of location by allowing auxiliary functions to be sublocations referring to a position within a complex value bounded to another location. Both locations in the sense of the standard ASMs and their sublocations are defined to be partial locations in this dissertation. Based on this extended view of partial locations, the dependency relation among partial locations has to be taken into account.

I specify data manipulations in a database transformation in terms of two kinds of partial updates on partial locations: shared updates and exclusive updates. Then the problem of partial updates can be circumvented by permiting only sets of partial updates which are consistent. However, having dependency relations among partial locations imposes some difficulties in the consistency checking of partial updates. The standard definition for the consistency of an update set has to be revised, as two partial updates with distinct but non-disjoint locations may still be conflicting.

The contribution of this chapter is the development of a systematic approach for consistency checking of partial updates on partial locations. This approach consists of two stages, first handling shared updates and second dealing with exclusive updates.

- *Stage One - normalisation of shared updates*

  This stage checks the operator-compatibility among shared updates. To improve the performance of database transformations, it is highly desirable to check the operator-compatibility of shared updates by utilising the schema information rather than examining possible database instances. For this reason, I propose an algebraic approach that exploits the compatibility of operators in $Opt(\ddot{\Delta}_\ell) = \{\mu | (\ell, b, \mu) \in \ddot{\Delta}_\ell\}$ for a multiset $\ddot{\Delta}_\ell$ of shared updates to the same location ℓ. The following result is obtained.

  Theorem 6.4.1 A non-empty multiset $\ddot{\Delta}_\ell$ of shared updates on the same location ℓ is operator-compatible if either $|\ddot{\Delta}_\ell| = 1$ holds, or there exists a $\mu \in Opt(\ddot{\Delta}_\ell)$ such that, for all $\mu_1 \in Opt(\ddot{\Delta}_\ell)$, $\mu_1 \preceq \mu$ (i.e., μ_1 is compatible to μ) holds.

  If $\ddot{\Delta}_\ell$ is operator-compatible, then it can be normalised into an update set $norm(\ddot{\Delta})$ containing only exclusive updates. This provides us with input for the second stage.

- *Stage two - integration of exclusive updates*

  This stage checks the compatibility of clusters of updates. Starting from a normalised update set, a family of clusters may be obtained by partitioning updates based on the dependence and subsumption relations among locations. Every cluster is said to be below some location ℓ in which ℓ is a location in the cluster subsuming every location in that cluster. For each cluster below a location ℓ, its consistency can be determined according to the following Theorem.

Theorem 6.5.1 Let Δ^{ℓ} be a cluster below the location ℓ. If Δ^{ℓ} is "level-by-level" value-compatible, then Δ^{ℓ} is consistent.

Chapter 7 provides a brief summary and assessment of the main results presented in this dissertation, and discusses some problems which are of interest for the future work.

Chapter 2

Literature Review

This chapter presents an overview of the related work. We start with a review of significant research findings in the area of database transformations. Several completeness criteria proposed in the literature for expressing different classes of database transformations are outlined. After that, we introduce the theory of Abstract State Machines (ASMs) with a focus on various characterisation theorems defined for different classes of algorithms and their representation by ASMs. In order to present a general picture of the development of generic computation models in database theory, we examine a variety of abstract machines tailored for database computations. In particular, some possible extensions for structures beyond finiteness and the connections between expressibility and logics are highlighted.

2.1 Database Transformations

In database theory, the study of database transformations dates back to relational database queries which originate from the seminal paper of Codd [57]. In this and later papers [56, 58, 59], a theoretical foundation for relational databases is established. Codd proposed a relational model of data in which a database is treated as a finite collection of finite structures. He also suggested two fundamental query languages for relational databases: relational calculus and relational algebra. Codd's theorem states that relational algebra and domain-independent relational calculus queries are precisely equivalent in expressive power, as both capture exactly the class of first-order queries over relational structures. A query language is called *relationally complete* if it can express all first-order queries.

Characterising relational database queries in a way independent from any specific language, Bancilhon [19] and Paredaens [117] independently discovered a criterion for completeness of relational query languages commonly known as BP-completeness. A relational query language is called *BP-complete* if the following conditions are satisfied for any input-output pair (S_{in}, S_{out}) of relational structures of a query:

- $Adom(S_{out}) \subseteq Adom(S_{in})$, and
- $Aut(S_{in}) \subseteq Aut(S_{out})$;

where $Adom(S)$ denotes the active domain of S (i.e., the set of constants appearing in S), and $Aut(S)$ denotes the automorphism group of S (i.e., the group of all functions from S to itself which preserve the structure of S).

The second condition requires the preservation of symmetry of the input in the output. This condition and the principles that Aho and Ullman [15] considered a query language must obey were generalised to invariants under isomorphisms by Chandra and Harel [52]. Invariants under isomorphisms exactly captures the abstract nature of computations in database applications by dealing with the logical properties rather than the interpretations of domain elements. This has led to the well-known *genericity principle* in database theory [98].

Apart from preserving database isomorphisms, Chandra and Harel [52] extended the criterion of BP-completeness to the class of *computable queries* by treating queries as partial recursive functions between finite, relational structures. Their notion of *completeness* remedied the deficiency of Codd's idea of relational completeness whose inability to express the transitive closure property was noted by Aho and Ullman in [15]. In order to demonstrate their proposed notion of completeness for relational query languages, Chandra and Harel [52] developed a simple but complete programming language (called QL) that can be thought of as relational algebra augmented with the power of iteration. This emerged the first computationally complete query language. The most interesting part of the language is a mechanism for providing the power of iteration. A straightforward addition of an iterative construct such as *least fixpoint* or *while* operator to the relational algebra or calculus still leads to queries in PSPACE and does not suffice to simulate unbounded space on a tape of Turing machines [99, 152]. The approach Chandra and Harel [52] adopted in the language QL is to utilise *unranked relation* variables that are able to simulate counters.

The results of Chandra and Harel have inspired two branches of research into the completeness of database transformations. The first branch concentrates on exploiting the expressive power of database transformations by generalising the completeness criteria to object-creating or update languages, in which the genericity principle used by Chandra and Harel for relational query languages has to be re-considered due to the presence of newly created objects in object-creating languages or explicitly specified values in update languages. The second branch focuses on understanding the expressibility of database transformations, that is, their completeness in terms of logics and their computational complexity. The paper by Van den Bussche et al. [150] provides a good survey of research results in the first branch. In the following, we will briefly introduce main results concerning the completeness criteria. As for achievements in the second branch, we will present an overview in Section 2.3.

In contrast to the mature and elegant theory of relational database queries resulting from extensive studies over the years, the understanding of database updates in relational databases is paltry. As a first step towards a formal investigation of database updates, Abiteboul and Vianu [9] initiated a study on the relationship between update languages and existing query languages. A notion of *update completeness* was proposed, which is not directly comparable but closely related to completeness of query languages as defined by Chandra and Harel [52]. Both deterministic and non-deterministic updates were considered from a transactional point of view. In order to deal with constants explicitly introduced by updates which may or may not be in the active domain of a database, the genericity property was relaxed to allow for the interpretation of a finite number of constants. A weaker version of genericity called *C-genericity* was introduced, which requires invariants under isomorphisms that fix a finite set C of constants.

In database theory, the formal usage of database transformations with binary relations on database structures defined by database programs first appeared in Abiteboul and Vianu's technical report [10]. With the term of *database transformation*, Abiteboul and Vianu on the one hand unified database queries and updates under one umbrella, and on the other hand widened the relational view for databases to complex-value databases. In the papers [11, 12], Abiteboul and Vianu employed the mechanism of *value invention* to introduce new domain elements, with which they were able to propose two complete languages in the sense of Chandra and Harel's completeness criterion: a procedural language in [11] and a logical programming language in [12]. In the case of deterministic database transformations, new invented values are only allowed to occur in "temporary" relations of a computation. This means, for any input-output pair (S_{in}, S_{out}) of instances associated with a deterministic database transformation the condition $Adom(S_{out}) \subseteq Adom(S_{in})$ holds.

Shortly after that, Abiteboul and Kanellakis [2] also applied the mechanism of value invention for the study of the object-oriented paradigm. The model they developed consists of a structural part that generalised many previously introduced complex-value data models, and an operational part (called IQL) that was shown to contain many popular, declarative query formalisms. In connection with set manipulations, the use of value invention appeared to be an intuitive way of creating object identifiers, which becomes the essential part of representing and manipulating complex structures in object-based databases. Nevertheless, the presence of new objects in an output instance gives rise to the issue of non-deterministic choice over newly created object identifiers. Therefore, the completeness of Chandra and Harel needed to be re-formulated for the object-oriented paradigm. Abiteboul and Kanellakis [2] came up with the solution to formulate the concept of *determinate transformation* by specifying conditions that a binary relation on database structures should satisfy, in order to qualify as a database transformation. These conditions described some desirable properties of database transformations: *well-typedness*, *effective computability*, *genericity* and *functionality* respectively. A determinate transformation is indeed a non-deterministic transformation that has output instances equivalent up to the renaming of new domain elements. Apart from the extension to allow the presence of object identifiers that were not in the input instance in the output instances, object identifiers were considered as atomic elements and only their interrelationships matter.

IQL is complete in the sense of Chandra and Harel's completeness criterion. Unfortunately, IQL is not complete for determinate transformations unless a construct dealing with copy elimination is explicitly added to the language [2, 63]. Motivated by this unsatisfactory situation, the class of *constructive transformations* was discovered by Van den Bussche [146, 150]. The characterisation of constructive transformations simply requires that all input-output pairs of instances associated with a determinate transformation must satisfy a condition formulated by Andries and Paredaens in [17]. This condition generalises the condition of Bancilhon [19] and Paredaens [117] for the relational structures to the context of structures with object creation. More specifically, for an input-output pair (S_{in}, S_{out}) of instances such that $Adom(S_{in}) \subseteq Adom(S_{out})$, there must exist an extension homomorphism from $Aut(S_{in})$ to $Aut(S_{out})$. In [150] Van den Bussche et al. showed that the expressiveness of the database transformation languages GOOD [91], IQL [2, 3] and FO+**new**+**while** (a minimal language defined as the closure of first-order logic under unbounded looping and associating new domain elements to tuples and sets of values) are equivalent, and complete with respect to constructive transformations. Furthermore, they established a close correspondence between object creation and the construction of hereditarily finite sets over the domain elements. This reconciles two different views for value invention: the pure object-creation perspective and the perspective of treating complex objects as unbounded value structures built up by recursively applying set and tuple constructors [62, 96, 97].

In contrast to the class of constructive transformations which further restricts the determinacy criterion of Abiteboul and Kanellakis, Van den Bussche [146, 148] also introduced the class of *semi-deterministic transformations* via a relaxation of the criterion. A semi-deterministic transformation is a non-deterministic transformation that has isomorphic output instances via isomorphisms that are automorphisms of the input instance. Augmenting IQL with a choice operator such as *witness* in [2, 148] or *swap-choice* in [92], the resulting language is complete in terms of the criterion of semi-deterministic transformations. In [149] Van den Bussche and Van Gucht tried to extend the concept of semi-determinism from queries to arbitrarily updates; however, they were hindered by several problems, such as: (1) how the composition of two semi-deterministic transformations should be defined; (2) what a language that is complete for semi-deterministic transformations should look like. Thus, they concluded that queries and updates may have some fundamental distinctions. The question of whether there exists a framework unifying both queries and updates has since been open.

The goal of this thesis is to investigate this problem from an algorithmic point of view. More specifically, database transformations are considered in terms of algorithms rather than merely computable functions. By employing the method of Abstract State Machines (ASMs) we open the way to unifying queries and updates in an elegant theoretical framework. In the next section, we will present some interesting research results relating to characterisation theorems for different notions of algorithms and their simulations by a variant of Abstract State Machines.

With the rise of the Extensible Markup Language (XML) as a standard for data exchange on the World Wide Web, XQuery [40] is emerging as a de-facto standard for querying XML data. Nevertheless, to provide sufficient support for Web-based applications in a service-oriented computing environment, flexible interactions between queries and updates over XML documents are highly demanded. As XML documents are commonly regarded as trees, it becomes important to study the properties of database transformations over tree structures. Over the last decade, a lot of research effort has been put in this area [6, 42, 53, 64, 121, 24]. However, the emphasis was usually on querying, while updates were neglected. Several extensions of XQuery to encompass updates that we are aware of [23, 50, 51, 70, 137] generally use explicit *snapshot semantics* to control the evaluation order of updates at certain level of snapshot granularity, which are often accompanied with some restrictions on the expression usage and the design of error handling. It turns out that a well-founded theory with high-level specification for XML database transformations including updates is still missing. In terms of object-creation results in tree structures, Van den Bussche made an observation in [147] which asserts that tree nodes that are not necessarily existing in an input database can be constructed via node construction expressions of an XML transformation language, in which case determinate and constructive transformations capture exactly the same class of database transformations.

The structures of a collection of related XML documents may be constrained by the presence of some XML schema, and such an XML schema acts as types imposed on XML database transformations. In the literature, there are several formalisms for describing schema information of an XML document, e.g., Document Type Definition (DTD) and its variations [16, 116], XML Schema [25, 142], RELAX NG [55, 54], etc. The paper [16] provides a good survey for various restrictions and extensions on DTD, and the papers [104, 111] provide an extensive coverage for an overall comparison of some popular schema formalisms.

2.2 ASMs and the Characterisation of Algorithms

Abstract State Machines (ASMs), formerly called *evolving algebra*, were introduced by Gurevich [78]. The formal semantics of ASMs was presented in [80, 81]. The initial intention for inventing ASMs was to develop an analogue to Church's thesis for formalising algorithms at a flexible abstraction level. The capability to model algorithms at different abstraction levels distinguishes ASMs from other computation models (such as, Turing Machines [143], Schönhage storage modification machines [128], Kolmogorov-Uspensky machines [103], etc), which are usually fixed to a certain level of abstraction. As a simple but powerful specification method, ASMs is supported by both a well-founded theoretical foundations and extensive applications in various areas of computer science. The universality and mathematical rigor of computation models described by ASMs have been extensively discussed in [27, 32, 47, 82, 119, 134] and verified by a large number of industry applications, such as, hardware and software architectures [71, 113], programming languages [135], protocols and security [22, 122] - just to mention a few. For more information on historical surveys and bibliographies of ASMs, the reader may refer to the ASM web site

[94]. In brief, the advent of ASM sheds light on establishing a bridge between theoretical computation models and practical system engineering methodology. In this section we will focus on providing an overview for characterising algorithms in the context of ASMs.

It was in [80] that Gurevich conjectured that ASMs can simulate any algorithm in lock-step at any arbitrary abstraction level. States of an algorithm are abstracted to first-order structures, independent from any specific representations. In addition to sequential algorithms and their extensions, the tentative definitions for non-sequential computations such as parallel and distributed algorithms were also provided. More specifically, both sequential and parallel algorithms are single-agent computations moving from an initial state to its next state, and go on until a final state is reached. Nevertheless, each step of sequential algorithms allows only a bounded amount of work to be done, whereas parallel algorithms are free from such restriction. This means that a single step of parallel algorithms may involve a number of invisible auxiliary agents. In contrast, distributed algorithms are considered to be multi-agent computations with partially ordered runs [84]. Agents in a distributed algorithm can share functions with each other and also have co-operative actions. It has been proven that ASMs and their extensions formalise certain notions of "algorithm", e.g., sequential and parallel algorithms have been proved in the papers [82] and [27] respectively. A restricted but important class of "small-step" distributed algorithms was proven in [72]. However, the characterisation for capturing more general distributed algorithms still remains a challenging problem.

The sequential ASM thesis of Gurevich [82] established the characterisation theorem for sequential algorithms, i.e., every sequential algorithm can be simulated, step for step, by a sequential ASM and vice versa. Sequential algorithms are defined in terms of three postulates: *sequential time*, *abstract state* and *bounded-exploration*, which characterise sequential algorithms in full generality. Following this work, the sequential ASM thesis was extended to a number of characterisation theorems targeted at different classes of algorithms [27, 28, 30, 31, 34, 35, 90]. Similar to the sequential ASM thesis, it was asserted that every algorithm in a class is behaviourally equivalent to a variant ASM customised to such a class. These extensions take into account various issues that may possibly arise in different classes of algorithms, such as the interaction with the environment, the background of computations, and non-determinism. In the following, we present these issues in turn.

In accordance with the sequential time postulate, an algorithm computes in a sequence of discrete steps where each step is a transition between two states. For a deterministic algorithm, there exists at most one possible state transition from any state and the one-step transformations are represented as function mappings. A non-deterministic algorithm, in contrast, may have a set of successor states relating to a state during one-step transformations. In [90], a complete characterisation for bounded-choice sequential algorithms was provided. It permits a restricted form of non-determinism in which finitely many states were considered as successors of a state by the transition relation. To reflect differences in the non-deterministic version, the authors reformulated the sequential time, abstract state and bounded-exploration postulates into the *nondeterministic sequential time*, *nondeterministic abstract state* and *bounded choice* postulates, respectively.

Constrained by the bounded-exploration postulate, any single step of a sequential algorithm can only explore a finite number of elements in a state. This number is entirely bounded by the algorithm itself and thus independent from the state, the environment and so forth. In order to distinguish algorithms with this feature from other algorithms that may have more complicated work involved in a single step between two states, algorithms satisfying the bounded-exploration postulate were called *small-step algorithms* while algorithms that have no bound on the amount of work done during a single step were called *wide-step algorithms* [83]. The charactersation theorem for wide-step algorithms that exhibit unbounded parallelism but only bounded sequentiality within a single step is presented in [27](also called parallel algorithms in [80]); it is constituted of six postulates: *sequential time*, *abstract state*, *background*, *proclets*, *bounded sequentiality* and *updates*. The sequential time and abstract state postulates are the same as for the sequential ASM thesis. The proclets and bounded sequentiality postulates address what can be done within a wide step. More specifically, a wide step of a parallel algorithm consists of a hierarchy of sub-steps in which a sub-step can be split into an unbounded number of smaller steps and go on until the split steps are small-steps of certain sequential algorithms. Blass and Gurevich coined the term *proclet* for referring to such sequential algorithms with small-steps. Each proclet must satisfy the postulates for sequential algorithms, and work in a local state that consists of a global state and some specific part. All proclets execute the same sequential algorithm in possibly different local states; moreover, the times taken to split into sub-steps in the hierarchy of a wide-step is bounded entirely by the algorithm itself. The updates postulate states that the updates produced by all the proclets contribute to the update set leading to the next global state. The background postulate asserts that a richer background is needed for parallel algorithms, i.e., each state must contain multisets, ordered pairs, boolean values, etc., and standard functions to manipulate them and a closed term *Proclets* for the set of proclets. Most parallel algorithms satisfy these postulates, however, parallel algorithms with the creation of components on the fly cannot be captured by them. This flaw was repaired in [32] by liberalizing the prior postulates to allow proclets to create additional proclets on the fly.

The problem of partial updates was first observed by the group on Foundations of Software Engineering at Microsoft Research during the development of the executable specification language AsmL [85, 86] that is built upon the theory of ASMs. In a nutshell, the problem is caused by two factors: complex objects and parallel modifications. The framework developed in [27] for parallel algorithms eliminates *partial updates* by viewing them as messages to proclets that integrate partial updates into appropriate *total updates* as noted in Remark 10.4. of [27]. However, the problem of partial updates does arise as an indispensably interesting phenomenon of algorithms that involve a certain degree of parallelism. In [87], Gurevich and Tillmann conducted a formal investigation on the problem of partial updates over data types *counter*, *set* and *map*. An algebraic framework is established by defining *particles* to be unary operations over a datatype and the parallel composition of particles to be an abstraction of order-independent sequential composition. However, this treatment of the parallel composition of particles fails to address partial updates over the data type *sequence*, as exemplified in [88]. This limitation motivated Gurevich and Tillmann to search for a more general solution for the problem of partial updates. In [88, 89], Gurevich and Tillmann proposed applicative algebras to address the partial update problem in a general algebraic framework. It turned out that the approach used in [87] was a special kind of an applicative algebra. Moreover, the problem of partial updates over *sequences* and *labeled ordered trees* was shown to be solvable in this new framework.

Another issue arising in the characterisation of parallel algorithms concerns the background of computations. Each parallel algorithm must have a background containing multisets and ordered pairs of its members, etc., as stated in the background postulate [27]. In [36], ASMs with set-theoretic background were studied, in which a base set is expanded to include all hereditarily finite sets built from the atoms. However, backgrounds are not issues relating only to parallel algorithms. In general, any algorithms such as those occurring in programming languages are associated with a specific background, no matter whether they are sequential, parallel or distributed in the sense of the classification in [80]. As pointed out by Blass and Gurevich in [26, 29], the purpose of investigating background structures of computations is two-fold. One is to describe the specific environment of an algorithm, such as built-in data types, predefined elements or operations, etc. The other is to formalise the view of elements newly imported from the *reserve* of a state in a computation. Blass and Gurevich initialised the formal discussion on various backgrounds of algorithms in [26]. By using the notion of *background class*, Blass and Gurevich [26] captured the class of structures that can be constructed on top of atoms without putting any structure (except equality) on the atoms themselves in a particular background. As part of the motivation for seeking a general definition for background class, several common background classes in connection with constructors such as set, string, list, etc. were exemplified as well, (called set, string and list backgrounds, respectively). However, it was shown by Tatiana Yavorskaya that the definition for background class was not precise enough. As a result, Blass and Gurevich reviewed the problem of background classes in [29], in which they generalised the theory of background structures presented in [26]. It leads to a more general framework for background classes, in which several properties like *equivalence*, *finitary*, *constructivity*, etc., of background classes were further scrutinised.

By considering different sorts of interactions between algorithms and environments, the characterisation theorem of algorithms can also be extended. Algorithms interacting with environments are classified as interactive algorithms, and have been investigated in [28, 30, 31, 34, 35]. There were two points of interest when dealing with the interaction between an algorithm and its environment. The first issue is whether the interaction is *inter-step* or *intra-step*. Inter-step means that algorithms can interact with their environments only between steps, while intra-step means that algorithms can also interact with their environments during a step. The second issue is whether the interactions are *ordinary*. An ordinary algorithm has the following features: (1) a step (either small or wide) will not finish until all queries from that step have been answered; (2) the information from the environment cannot be used except for answers to queries. The papers [28, 30, 31] formulate the postulates, a variant of abstract state machines and the characterisation theorem for ordinary, interactive, small-step algorithms, respectively. The overall result is that ordinary, interactive, small-step ASMs were shown to be behaviourally equivalent to ordinary, interactive, small-step algorithms. This work has been further generalised to arbitrary interactive, small-step algorithms in [33, 34, 35]. The papers [18, 83] give a systematic discussion of a variety of interactive algorithms that have been studied in the literature.

2.3 Generic Computation Models in Database Theory

To investigate queries and updates over abstract domains, several generic computation models have been developed in database theory. In this section, we briefly present some of the research results related to their expressibility in terms of logics, and their perspectives for the underlying structures.

2.3.1 Abstract Machines

The study of abstract machines in relation to databases was initiated by Abiteboul and Vianu [13]. A generic computation model, called *relational machine*, was developed to address the mismatch between the generic nature of database computations and the classical complexity measures based on Turing Machines. Augmented with a *relational store*, i.e., a finite set of fixed-arity relations, a Turing machine is extended to a relational machine in which the relational store can only be modified via a first-order query. Therefore, relational machines are in fact computation devices that embed a relational database query language in a host programming language. Two versions of relational machines were discussed by Abiteboul and Vianu [13], differing only in the interactions between the Turing tape and the relational store. It was shown that the weaker version (called GM^{loose}) is a powerful extension of many well-known query languages, such as fixpoint and while queries, but not computationally complete. For example, the parity of an input set cannot be computed by GM^{loose}. The strong version (called GM) is complete with the involvement of parallelism, and the output can only be generated after all results from subcomputations have been integrated to ensure the genericity of the computation. Based on relational machines, computability or complexity of relational database languages can be analyzed over abstract structures without any presumed order over database domains. In contrast to Turing Machines that operate on a sequential encoding of structures, relational machines handle abstract structures in a generic way. This leads to the *discerning power* of relational machines, i.e., the power to distinguish between different pieces of their input [8]. Therefore, Turing machines may naturally measure complexity in terms of the size of such an encoded input, but relational machines can only measure the size of an input with respect to their discerning powers. Several complexity classes for generic computations were defined as generic analogy of classical complexity classes. An infinitary logic was related to relational machines in the paper [7]. It turns out that relational machines correspond to the natural effective fragment of infinitary logic, and other well-known query languages are related to infinitary logic using syntactic restrictions formulated in language-theoretic terms. The relationship between logics, relational machines and complexity was further studied in [8, 14].

Generalising relational machines with reflection, Abiteboul et al. [4, 5] suggested the use of *reflective relational machine* to deal with the dynamic generation of queries in a host programming language. A query in reflective relational machines can be constructed (partially or fully) by a program rather than a programmer. In essence, a reflective relational machine consists of a Turing machine component operating sequentially and a relational computation counted as one parallel step. Abiteboul et al. [4, 5] showed that reflective relational machines capture parallel models of computation from a practical point of view, and the reflection allows machines to express more parallelism than non-reflective ones. Subsequently, the connections between reflective relational machines and parallel complexity classes were addressed in [4]. In the paper [14] it was shown that in terms of a fixed number of variables allowed on the Turing tape the relational machine cannot obtain additional power for the augmentation with reflection. This line of research was continued in [4, 5], which investigate the expressive power of reflective machines when the number of variables that can be used is restricted. Their study revealed that the power of reflection relies on having an unbounded number of variables used in queries.

Taking into account numerical queries that have values queried over natural numbers, there is another direction to extend the power of (reflective) relational machines. One tricky issue relating to numerical queries is that numerical values occurring in computations are not necessarily in the domain of the given structure. This can lead to structures of computation going beyond finiteness. In order to take the computation of numerical queries into account, Torres [144] defined an extension of the reflective relational machine, called *untyped reflective relational machine*, and proved in [52] that the extended relational machines are complete in terms of both *typed and untyped queries*, i.e., queries of fixed arity and of unfixed arity, respectively. In another paper [145], Torres further developed the notion of *reflective counting machine* based on reflective relational machines to investigate a hierarchy of computable queries over relational databases, in terms of the preservation of equality of theories in fragments of the first-order logic with bounded number of variables and counting quantifiers.

In order to investigate the computability of queries in the presence of external functions, relational machines have also been extended for databases with nested relations and external functions, called *relational machines for complex objects* [136]. The result showed that relational machines are complete on nested relations, although they are known to be incomplete on flat relations.

2.3.2 Expressibility and Logics

Compared with classical computation models like Turing machines, the most significant contribution of (reflective) relational machines is that they are computation devices operating directly on abstract structures rather than encodings of structures. This generic viewpoint for database computations coincides with the logical view for computations over unordered structures. Driven by the interest of investigating a logic that captures a natural fragment of Polynomial-time (PTime) algorithmic problems over abstract structures without the presence of a linear order, Blass et al. [36] developed a computation model over relational databases which was later called *BGS model* in [38]. The BGS model is a variant of ASMs with a universe containing all hereditarily finite sets built from the atoms. Blass et al. [36] attempted to capture the choiceless fragment of PTime algorithmic problems by replacing arbitrary choice with parallel execution in the BGS model that do not distinguish between isomorphic structures. The resulting logic, called *Choiceless Polynomial Time logic* ($\widetilde{C}$PT), still does not capture all PTime algorithmic problems but has been shown to be strictly more expressive than the fixed-point logic in which PTime computations over linearly ordered structures are definable [99, 152]. Continuing with the consideration of polynomial-time computations over unordered structures, the same authors further studied in [37] an extension of the logic $\widetilde{C}$PT because they had observed that $\widetilde{C}$PT could not express cardinality properties such as counting of or computing the evenness of the number of elements in a set. They added counting terms into the logic. By examining several algorithmic problems residing at the border of the known complexity classes contained in PTime algorithmic problems [37], the extended logic (called $\widetilde{C}$PT+Card) is found to be strictly more expressive than the logic developed by Cai et al. [49], which is an extension of the fixed-point logic with counting terms being able to express cardinality properties. The connections between the BGS model and other computational complete query languages such as QL [52] and $while_{new}$ [1] are discussed in-depth in [38]. Their results showed that the BGS model is powerful enough to express all computable queries in relational databases. In terms of polynomial-time computations, BGS is strictly more powerful than QL and $while_{new}$. A language extending $while_{new}$ by set-based invention and BGS can simulate each other with only linear-step, polynomial-space overhead. According to the results from [13, 150] that IQL is polynomial-time equivalent to $while_{new}$, it follows that IQL and BGS are also PTime equivalent.

To extend the application of logics from queries to updates, a number of logical formalisms have been developed to provide the reasoning for both states and state changes in a computation model [45, 151]. A popular approach was to take dynamic logic as a starting point and then to define the declarative semantics for logical formulae based on Kripke structures. It led to the development of the *database dynamic logic* (DDL) and *propositional database dynamic logic* (PDDL) by Spruit et al. [130, 132, 131]. DDL has atomic updates for inserting, deleting and updating tuples in predicates and for functions, whereas PDDL has two kinds of atomic updates: passive and active updates. Passive updates change the truth value of an atom while active updates compute derived updates using a logic program. In [133] Spruit et al. further proposed *regular first-order update logic* (FUL), which is a generalised version of dynamic logic tailored for specification of database updates. A state of FUL is viewed as a set of non-modal formulae. Unlike standard dynamic logic, predicate and function symbols rather than variables are updateable in FUL. Two instantiations of FUL were also discussed by the authors. One is called *relational algebra update logic* (RAUL) which is an extension of relational algebra with assignments as atomic updates. Another is DDL which can be obtained by parameterizing FUL by two kinds of atomic updates: bulk updates to predicates and assignment updates to functions. It was shown that DDL is also "update complete" in terms of the update completeness criterion proposed by Abiteboul and Vianu [9].

The emergence of *transaction logic* [43, 44, 46] manifests a quite different but very elegant approach for dealing with the dynamics of database computations. Transaction logic is an extension of classical predicate logic by adding operators *serial conjunction*, *parallel conjunction*, etc. to account for transactional features. At the same time, transaction logic retains the oracles for data and transition, i.e., there are no any assumptions about states and updates on states of a database. It was also shown that transaction logic has a sound and complete proof theory. As transaction logic adopts a transactional perspective, and is therefore powerful in specifying database transactions at many levels of detail. In doing so, transaction logic turns out to be a general logic for database transactions that is able to take care of all kinds of transactional features. Nevertheless, it is not a logic dedicated to characterising the effect of updates on states, which is at the core of database transformations as described in Section 2.1.

As we alluded to before, ASMs turn out to be a promising approach for the specification of database transformations. The logical foundations for ASMs have been well studied from several perspectives [76, 77, 120, 127, 134]. Groenboom and Renardel de Lavalette presented a logic called *modal logic of creation and modication* (MLCM) in [76], which is a multimodal predicate logic intended to capture the ideas behind ASMs. On the basis of MLCM the authors also developed a language called *formal language for evolving algebras* (FLEA) in [77]. Instead of values of variables, states of an MLCM are represented by mathematical structures expressed in terms of dynamic functions. Renardel de Lavalette generalise in [120] the logic MLCM and other variations from [69] to *modification and creation logic* (MCL) for which there exists a sound and complete axiomatisation. In [127], an extension of dynamic logic with update of functions, extension of universes and simultaneous execution(called EDL) is presented, which allows statements about ASMs to be directly represented. A logic complete for *hierarchical ASMs* i.e., ASMs that do not contain recursive rule definitions, was developed by Stärk and Nanchen [134]. This logic differs from other logics for ASMs in two respects: (1) the consistency of updates has been accounted for in the reasoning part of the logic; (2) modal operators are allowed to be eliminated in certain cases. Nanchen [112] also applied this logic for the verification of access and update properties of a certain class of ASMs. In this dissertation, we will extend the logic for hierarchical ASMs towards the characterisation of database transformations, which allow a limited form of non-determinism to be captured.

2.3.3 Beyond Finiteness of Structures

In relational database theory, structures of a database computation have predominantly been regarded as being finite, and deeply studied in the field of finite model theory [68, 79]. The typical example is relational databases. In association with a set $\{D_i\}_{i\in[1,n]}$ of domains, a relational database consists of a finite set of relations defined to be the subsets of cartesian products of some domains, i.e., $R \subseteq D_1 \times ... \times D_k$ for $k \leq n$. Although domains maybe countably infinite, only a finite number of elements occur in the finite number of relations of a relational database. By ignoring elements of infinite domains that do not occur in any relations, a relational database is in fact a finite structure.

The view of databases as finite structures greatly inspires the development of finite model theory. Fruitful results on the relationship between logical definability and computational complexity over finite structures were found in [73]. Nevertheless, when handling algorithmic issues that often arise from applying algorithmic toolboxes in database-related problems, the finiteness condition on database structures turns out to be too restrictive. There are several database applications in which certain kind of infinity is indispensable.

Firstly, database computations over finite structures may deal with new elements from countably infinite domains, outside of the original finite structure. For instance, counting queries produce natural numbers even if no natural number occurs in a finite structure, and aggregation and arithmetic operations can generate elements that are not the elements of a finite structure. Moreover, finite structures may have invariant properties which possibly have infinite elements implied in satisfying them, such as, numerical invariants of geometric objects or database constraints. Last but by no means least, every database computation either implicitly or explicitly lives in a certain background that supplies all necessary information relating to the computation and usually exists in the form of infinite structures.

To remedy such deficiencies, several extensions on finite structures have been suggested in the literature (e.g., [74, 52, 75]). A common idea among these extensions is to treat finite structures as the primary part, and to provide extra finite or infinite structures as the secondary part. In order to relate the primary and secondary parts together, several kinds of terms, formulae or functions are defined. More specifically, it was Chandra and Harel [52] who first observed limitations of finite structures in database theory. They proposed a notion of an *extended database* which extends finite structures by adding another countable, enumerable domain containing interpreted features such as numbers, strings and so forth. The main intention of their study was to provide a more general framework that can capture queries with interpreted elements but no further study was carried on to study the internal part of the added domain. Another kind of extension was largely driven by the efforts to solve the problem of expressing cardinality properties [49, 75, 100, 114, 115, 144, 145, 37]. For example, Grädel and Otto [75] developed a two-sorted structure which adjoins a one-sorted finite structure with an additional finite numerical domain, and adds the terms expressing cardinality properties. They aimed to study the expressive power of logical languages that involve induction with counting on such structures. A promising line of work is *meta-finite model theory* proposed by Grädel and Gurevich [74]. They formalised several variations of meta-finite structures, which all consist of: a primary part that is a finite structure, a secondary part that may be a finite or infinite structure, and a set of weight functions from the primary part into the secondary part. Among these variations, meta-finite structures with multiset operations are of most interest to us due to its naturalness and generality.

Chapter 3

Foundations of Database Transformations

According to [10] a database transformation is a binary relation on database instances that encompass queries and updates. In general, a database transformation can be non-deterministic, but it must be recursively enumerable and generic in the sense of preserving isomorphisms. Instead of viewing database transformations as computable functions, the focus of this chapter is to completely characterise the algorithms that transform input databases into output databases. It is analogous to the seminal work on the sequential ASM thesis [82].

This chapter has two tasks. The first one is to characterise database transformations by a set of simple and intuitive postulates. These postulates should cover all database transformations while leaving sufficient latitude to specify the specific characteristics of data models such as the relational, object-oriented, object-relational and semi-structured data models. We present five postulates along with a detailed explanation in Section 3.1. The second task is to develop a general computation model for database transformations. This is a crucial step towards establishing a unifying theoretical framework. In Section 3.2 a variant of ASMs is developed for database transformations, which we call *database Abstract State Machines* (DB-ASMs). We show that DB-ASMs satisfy the five postulates. This is relatively easy to achieve while the converse, that all computations stipulated by the five postulates for database transformations can be simulated by a behaviourally equivalent DB-ASM, is much harder. The proof for this result, that DB-ASMs capture all database transformations, is presented in Section 3.3.

3.1 Postulates for Database Transformations

We formally introduce the five postulates for database transformations, which are the *sequential time postulate*, the *abstract state postulate*, the *background postulate*, the *bounded exploration postulate*, and the *bounded non-determinism postulate*.

Definition 3.1.1. An object satisfying these five postulates is a *database transformation*.

3.1.1 Sequential Time

As in [27] and [82] a database transformation, the same as any algorithm, proceeds stepwise on a set of states. It starts somewhere, which gives us a set of initial states. Although it makes perfect sense to consider non-terminating algorithms, we will only consider database transformations which terminate somewhere, and thus associate with it a set of final states. As discussed in [82] this restriction is more a technicality as far as the work in this dissertation is concerned, but is necessary in light of our aim to embed the results into a theory of database systems, in which many database transformations have to co-exist and interplay with each other.

Deviating from the sequential time postulate in the sequential and parallel ASM theses, a one-step transition relation on states is used for database transformations, as in the characterisation of bounded-choice sequential algorithms [90], rather than a one-step transformation function on states. Doing so introduces non-determinism into database transformations. However, the extent of permissible non-determinism will be limited by further postulates. In fact, only non-deterministic choice among the finite answers to a query is permitted. Non-determinism in database transformations becomes important when we are faced with object creation in certain data models, as discussed intensively in [146, 148, 149].

Postulate 3.1.1 (sequential time postulate). A database transformation T is associated with a non-empty set of states $\mathcal{S}_T$ together with non-empty subsets $\mathcal{I}_T$ and $\mathcal{F}_T$ of initial and final states, respectively, and a one-step transition relation δ_T over $\mathcal{S}_T$, i.e., $\delta_T \subseteq \mathcal{S}_T \times \mathcal{S}_T$.

The sequential time postulate allows us to define the notion of a *run* as analogous to sequential and parallel algorithms. As termination is required, a run must be finite, ending at the first final state that is reached. Nevertheless, an initial state in a run is permitted to also be a final state. The motivation behind this is that database transformations are always associated with a database system, in which each run of a database transformation produces a state transition. The final state of a transition becomes the initial state for another transition, leading to an infinite sequence of states that result from running a set of database transformations in a serial (or serialisable) way. This view of database systems has been stressed in [107, 108, 109].

Definition 3.1.2. A *run* of a database transformation T is a finite sequence $S_0, \dots, S_f$ of states with $S_0 \in \mathcal{I}_T$, $S_f \in \mathcal{F}_T$, $S_i \notin \mathcal{F}_T$ for $0 < i < f$, and $(S_i, S_{i+1}) \in \delta_T$ for all $i = 0, \dots, f-1$.

Database transformations will be considered only up to behavioural equivalence.

Definition 3.1.3. Database transformations T_1 and T_2 are *behaviourally equivalent* if $\mathcal{S}_{T_1} = \mathcal{S}_{T_2}$, $\mathcal{I}_{T_1} = \mathcal{I}_{T_2}$, $\mathcal{F}_{T_1} = \mathcal{F}_{T_2}$ and $\delta_{T_1} = \delta_{T_2}$ hold.

In accordance with Definition 3.1.3, we know that behaviourally equivalent database transformations have the same runs.

3.1.2 Abstract States

The abstract state postulate is an adaptation of the corresponding postulate for sequential algorithms [82], according to which states are first-order structures, i.e. sets of (partial) functions, some of which may be marked as relational. These functions are interpretations of function symbols given by some signature. Following [82] we assume that each signature in a computation background (i.e., background signature as will be discussed in Subsection 3.1.3) contains the nullary symbols *true*, *false*, *undef*, a unary symbol *Bool*, the equality sign symbol and the symbols of the usual Boolean operations. With the exception of *undef* all these logic symbols are relational. Equality, truth values and Boolean operations are interpreted in a fixed way in all states.

Taking structures as states reflects a common practice in mathematics, where almost all theories are based on first-order structures. Variables are special cases of function symbols of arity 0, and constants are the same, but unchangeable. A function symbol is called *dynamic* if the function is changeable in states; otherwise, it is called *static*.

Definition 3.1.4. A *signature* Υ is a set of function symbols, each associated with a fixed arity. A *structure* over Υ consists of a set B, called the *base set* of the structure, together with interpretations of all function symbols in Υ, i.e. if $f \in \Upsilon$ has arity k, then it will be interpreted by a function from B^k to B.

Let S be a structure over Υ. For each ground term $t \in \Upsilon$ we use $val_S(t)$ to denote the interpretation of t in the structure S.

Definition 3.1.5. An *isomorphism* from structure S_X to structure S_Y is defined by a bijection $\varsigma : B_{S_X} \to B_{S_Y}$ between the base sets that extends to functions by $\varsigma(val_{S_X}(f(b_1, \dots, b_k))) = val_{S_Y}(f(\varsigma(b_1), \dots, \varsigma(b_k)))$. A *Z-isomorphism* for $Z \subseteq B_{S_X} \cap B_{S_Y}$ is an isomorphism ς from S_X to S_Y that fixes Z, i.e. $\varsigma(b) = b$ for all $b \in Z$.

As the base set B contains a value $\perp$ representing undefinedness, partial functions are captured by the undefinedness value $\perp$ associated with *undef*. Furthermore, relations are captured by letting $val_{S_X}(f(a_1, \dots, a_k)) = true$ mean that $(a_1, \dots, a_k)$ is in the relation f of S_X, and $val_{S_X}(f(a_1, \dots, a_k)) = false$ that it is not.

Z-isomorphisms are needed when dealing with constants in the base set that are represented by 0-ary function symbols. However, an automorphism ς of a structure S_X fixes all values $a \in B$ that are represented by ground terms. If $a = val_{S_X}(t)$ holds for some ground term t, then $\varsigma(a) = a$ holds for each automorphism ς of S_X. Thus, Z-isomorphisms can be neglected, as we could always add syntactic surrogates, i.e. 0-ary function symbols for all elements of Z to the signature Υ.

In the case of database transformations, two specific issues that affect the definition of states have to be taken care of.

The first database-specific issue is the intrinsic finiteness of databases. It is tempting to adopt finite model theory [66] and require that states for database transformations are finite structures. For instance, a state could be modelled by a finite set of finite relations [1] in the relational data model. This would not, however, capture the full picture. For example, a simple query counting the number of tuples in a relation would require natural numbers in the base set, and any restriction to a finite set would already rule out some database transformations. While finite structures avoid dealing with some odd problems, thus ensuring that query results are always finite, they put undesired limitations on the notion of states. In order to deal with this problem, we adopt the concept of *meta-finite structures* proposed by Grädel and Gurevich [74], henceforth called *meta-finite states*, in which actual database entries are treated merely as surrogates for real values, and consequently there will be three kinds of function symbols: those representing the database, those representing everything outside the database, and bridge functions that map surrogates in the database to values outside the database in possibly infinite domains. This permits the database to remain finite while allowing database entries to be interpreted in possibly infinite domains. Of particular importance is that we are able to generalise most of the results achieved in finite model theory to meta-finite models but avoid undesired complications coming from the infinity of the structure.

There is a subtlety here that needs to be addressed. In object-oriented databases identifiers for objects may be used. In tree-based databases identifiers for tree nodes may be required. Hence, in both cases there is a need to create new identifiers. As discussed in [82] the creation of new values is not a problem in principle. We can assume a set of reserve values from which these new identifiers can be taken. Nevertheless, it is not a-priori clear how many such values will be needed. This problem can be circumvented by requiring that only the active database domain is finite, that is, the set of elements from the base set appearing in the database part of a state.

The second database-specific issue is the presence of a data model that prescribes what a database should look like and what databases are meaningful. In the case of the relational model we would have to deal simply with relations, while in the cases of object-oriented and tree-based databases constructors for complex values such as finite sets, multisets, maps, arrays, union, trees and so on are needed. The consequences that this has on states will be separately addressed in Subsection 3.1.3 on the background postulate.

Definition 3.1.6. The *signature* Υ of a meta-finite structure is composed as a disjoint union consisting of a sub-signature Υ_{db} (called *database part*), a sub-signature Υ_a (called *algorithmic part*), and a finite set of bridge function symbols each with a fixed arity, i.e. $\Upsilon = \Upsilon_{db} \cup \Upsilon_a \cup \{f_1, \dots, f_\ell\}$. The *base set* of a meta-finite structure S is $B = B_{db}^{ext} \cup B_a$ with interpretation of function symbols in Υ_{db} and Υ_a over $B_{db} \subseteq B_{db}^{ext}$ and B_a, respectively, with B_{db} depending on the state S. The interpretation of a bridge function symbol of arity k defines a function from B_{db}^k to B_a. With respect to such states S the restriction to Υ_{db} is a finite structure, i.e. B_{db} is finite.

Postulate 3.1.2 (abstract state postulate)**.** Every database transformation T satisfies the following conditions:

- All states $S \in \mathcal{S}_T$ are meta-finite structures over the same signature Υ.
- Whenever $(S, S') \in \delta_T$ holds, the states S and S' have the same base set B.
- The sets $\mathcal{S}_T$, $\mathcal{I}_T$ and $\mathcal{F}_T$ are closed under isomorphisms, and for $(S_1, S_1') \in \delta_T$ each isomorphism ς from S_1 to S_2 is also an isomorphism from S_1' to $S_2' = \varsigma(S_1')$ with $(S_2, S_2') \in \delta_T$.

The abstract state postulate is an adaptation of the analogous postulate from [27, 82] to further consider states as being meta-finite structures, the presence of final states and the fact that one-step transition is a binary relation.

Example 3.1.1. Let us consider a database with the relational database schema shown in Example 1.1.1. From an abstract state point of view, we would get $B_{db} = D_n \cup D_c \cup D_b \cup \{true, false\}$ as a union of four disjoint, finite sets, and $B_a = \mathbb{R} \cup \Sigma$ as the union of the set $\mathbb{R}$ of rational numbers and the set Σ^* of character strings over an alphabet Σ, both infinite.

We would have the sub-signature Υ_{db} containing four function symbols: PUBLICATION, PERSON, ACADEMICUNIT and AUTHORSHIP. These function symbols are all relational and would be interpreted by

- PUBLICATION: $D_n \times D_c \times D_n \times D_c \times D_b \times D_c \times D_n \rightarrow \{true, false\}$,
- PERSON: $D_n \times D_c \times D_c \times D_c \rightarrow \{true, false\}$,
- ACADEMICUNIT: $D_n \times D_c \times D_n \times D_n \rightarrow \{true, false\}$,
- AUTHORSHIP: $D_n \times D_n \times D_n \times D_n \rightarrow \{true, false\}$.

The sub-signature Υ_a could contain function symbols for subtraction, addition, division, multiplication and others, for example, a binary function symbol $>$ for an ordering relation, and a unary function symbol EVEN for an even function. The function symbol EVEN would be interpreted by a function $\mathbb{N} \rightarrow$ {true, false} such that

$$\text{EVEN}(n) = true \text{ if n is even, and } false \text{ otherwise.}$$

As for the bridge functions, we would have the function symbols f_{num}, f_{char} and f_{blob} for the interpretation of elements in the domains D_n, D_c and D_b, respectively.

- f_{num}: $D_n \rightarrow \mathbb{N}$,
- f_{char}: $D_c \rightarrow \Sigma^*$,
- f_{blob}: $D_b \rightarrow \mathbb{R}$.

Using this representation of finite relations for PUBLICATION, PERSON, ACADEMICUNIT and AUTHORSHIP, a query such as "List all publications with an even number of authors" would be possible, provided we add a unary function symbol RESULT to the sub-signature Υ_{db} to pick up the result.

Similarly, a query such as "List the names of all persons whose publications have been downloaded the most number of times" would require unary function symbols MAX and COUNT in the algorithmic part of the signature. □

We adopt the idea of meta-finite states in the abstract state postulate, but we do not restrict the database part of a state to be relational. Consequently, we can capture data models other than the relational one.

Let us finally look at genericity as expressed by the preservation of isomorphisms in successor states. In the sequential ASM thesis sequential algorithms are considered to be deterministic, so a state S has a unique successor state $S^{'}$. It means that, by the abstract state postulate for sequential algorithms, an automorphism ς of S is also an automorphism of $S^{'}$. In our abstract state postulate for database transformations (i.e., Postulate 3.1.2), however, there can be more than one successor state of S, as δ_T is a relation. Now, if ς is an automorphism of S, and $(S, S') \in \delta_T$ holds, we obtain an isomorphism ς from S' to S'' where $S'' = \varsigma(S')$ with $(S, S'') \in \delta_T$. Thus, an automorphism of S induces a permutation of the successor states of S.

3.1.3 Backgrounds

The postulates 3.1.1 and 3.1.2 are in line with the sequential and parallel ASM theses [27, 82], and with the exception of allowing non-determinism in the sequential time postulate and the reference to meta-finite structures in the abstract state postulate there is nothing in these postulates that makes a big difference to postulates for sequential algorithms. The next postulate, however, is less obvious, as it refers to the background of a computation, which contains everything that is needed to perform the computation, but is not yet captured by states. For instance, truth values and their connectives, and a value $\perp$ to denote undefinedness constitute necessary elements in a background.

For database transformations, in particular, we have to capture constructs that are determined by the used data model, for example, relational, object-oriented, object-relational or semi-structured data models. That means, we will have to deal with type constructors and with functions defined on such type constructors. Furthermore, when we allow values (e.g. identifiers) to be created non-deterministically, we would like to take these values out of an infinite set of reserve values. Once created, these values become active, and we assume they can never be used again for this purpose.

Let us take the following example, which was used in [3] to illustrate a data model that generalises most of known complex object data models. In this model a distinction is made between abstract identifiers and constants. These elements stem from disjoint base domains, which together constitute the base set of database transformations. With the addition of constructors for tuples, sets, multisets, lists, et cetera, domains of arbitrarily nested complex-values can be built upon the base domains. For instance, a domain $D_{(Int,\{String\})}$ over base domains *Int* and *String* would represent complex-values consisting of an integer and a set of strings.

Example 3.1.2. Suppose that the state of a database has a universe containing abstract identifiers from domains $D_1 = \{o_{eve}, o_{adam}\}$, $D_2 = \{o_{cain}, o_{abel}, o_{seth}, o_{other}\}$, $D_3 = \{o_{n_k} | k = 1, ..., 5\}$, $D_4 = \{o_{m_k} | k = 1, ..., 3\}$, $D_5 = \{o_{d_1}, o_{d_2}\}$ and constants from domains *String* and *Bool*. Furthermore, assume constructors $\{\cdot\}$ for finite sets with unfixed arity, tuples $(\cdot)$ with arity up to 3, and union $\sqcup$ with arity 2.

Let the signature of states have function names *1st-generation*, *2nd-generation*, *name*, *occupation*, *descendant*, and relation names *founded-lineage*, *ancestor-of-celebrity* such that

- *1st-generation*: $D_1 \rightarrow D_{(nam:D_3, spou:D_1, children:\{D_2\})}$,
- *2nd-generation*: $D_2 \rightarrow D_{(nam:D_3, occu:D_4)}$,
- *founded-lineage*: 2nd-gen: $D_2 \rightarrow$ *Bool*,
- *ancestor-of-celebrity*: anc: $D_2 \times$ desc: $D_5 \rightarrow$ *Bool*,
- *name*: $D_3 \rightarrow D_{String}$,
- *occupation*: $D_4 \rightarrow D_{\{String\}}$,
- *descendant*: $D_5 \rightarrow D_{String \sqcup (spou:String)}$.

The interpretation of function and relation names is as follows:

- for *1st-generation*, $o_{eve} \mapsto$ (nam: o_{n_1}, spou: o_{adam}, children: $\{o_{cain}, o_{abel}, o_{seth}, o_{other}\}$) and $o_{adam} \mapsto$ (nam: o_{n_2}, spou: o_{eve}, children: $\{o_{cain}, o_{abel}, o_{seth}, o_{other}\}$),
- for *2nd-generation*, $o_{cain} \mapsto$ (nam: o_{n_3}, occu: o_{m_1}), $o_{seth} \mapsto$ (nam: o_{n_4}, occu: o_{m_2}) and $i_{abel} \mapsto$ (nam: o_{n_5}, occu: o_{m_3}),

- for *founded-lineage*, it is $\{$(2nd-gen: o_{cain}), (2nd-gen: o_{seth}), (2nd-gen: o_{other})$\}$,
- for *ancestor-of-celebrity*, it is $\{$(anc: o_{seth}, desc: o_{d_1}), (anc: o_{cain}, desc: o_{d_2})$\}$,
- for *name*, $o_{n_1} \mapsto$ Eve, $o_{n_2} \mapsto$ Adam, $o_{n_3} \mapsto$ Cain, $o_{n_4} \mapsto$ Seth and $o_{n_5} \mapsto$ Abel,
- for *occupation*, $o_{m_1} \mapsto$ $\{$Farmer, Nomad, Artisan$\}$, $o_{m_2} \mapsto \{\}$ and $o_{m_3} \mapsto$ $\{$Shepherd$\}$,
- for *descendant*, $o_{d_1} \mapsto$ Noah and $o_{d_2} \mapsto$ (spou: Ada).

That is, objects of *1st-generation* are described by a name, a reference to a spouse, and a set of references to children. Objects of *2nd-generation* are described by a name and a set of professions, and objects of *descendant* are described by a name only or a tuple with a name. The *founded-lineage* defines a set of objects of *2nd-generation*, and *ancestor-of-celebrity* is a simple binary relation.

Suppose we want to create a new object with a new identifier in D_3. For this, we obtain a new identifier $o_3 \in D_3$, from the set of reserve values. Then we set $name(o_3) :=$ Isaac and $ancestor\text{-}of\text{-}celebrity(o_{seth}, o_3) := true$ to update *name* and *ancestor-of-celebrity* (taking $false$ as the default value for all other cases). Similarly, with $founded\text{-}lineage(o_{other}) := false$ we would delete o_{other} from the *founded-lineage* relation. □

Following [27] we use background classes to define structures provided by backgrounds, which will then become part of states. Background classes themselves are determined by background signatures that consist of constructor symbols and function symbols. Function symbols are associated with a fixed arity as in Definition 3.1.4, but for constructor symbols we permit the arity to be unfixed or bounded.

Definition 3.1.7. Let $\mathcal{D}$ be a set of base domains and Υ_K a background signature. Then a *background class* K with Υ_K over $\mathcal{D}$ is constituted by

- the universe $U = \bigcup_{D \in \mathcal{D}'} D$ of elements, where $\mathcal{D}'$ is the smallest set with $\mathcal{D} \subseteq \mathcal{D}'$ satisfying the following properties for each constructor symbol $\llcorner\lrcorner \in \Upsilon_K$:
  - if $\llcorner\lrcorner \in \Upsilon_K$ has unfixed arity, then
    - $\llcorner D \lrcorner \in \mathcal{D}'$ for all $D \in \mathcal{D}'$, and $\llcorner a_1, \dots, a_m \lrcorner \in \llcorner D \lrcorner$ for every $m \in \mathbb{N}$ and $a_1, \dots, a_m \in D$, and
    - $A_{\llcorner\lrcorner} \in \mathcal{D}'$ with $A_{\llcorner\lrcorner} = \bigcup_{\llcorner D \lrcorner \in \mathcal{D}'} \llcorner D \lrcorner$;

- if $\llcorner\lrcorner \in \Upsilon_K$ has bounded arity n, then $\llcorner D_1, \ldots, D_m \lrcorner \in \mathcal{D}'$ for all $m \leq n$ and $D_i \in \mathcal{D}'$ $(1 \leq i \leq m)$, and $\llcorner a_1, \ldots, a_m \lrcorner \in \llcorner D_1, \ldots, D_m \lrcorner$ for every $m \in \mathbb{N}$ and $a_1, \ldots, a_m \in D$;
- if $\llcorner\lrcorner \in \Upsilon_K$ has fixed arity n, then $\llcorner D_1, \ldots, D_n \lrcorner \in \mathcal{D}'$ for all $D_i \in \mathcal{D}'$ and $\llcorner a_1, \ldots, a_n \lrcorner \in \llcorner D_1, \ldots, D_n \lrcorner$ for all $a_i \in D_i$ $(1 \leq i \leq n)$,

- and an interpretation of function symbols in Υ_K over U.

To help a better understanding of the notion background class, we provide several examples in the following.

Example 3.1.3. Let us consider the BGS model proposed in [38]. The BGS model has elements from the domain D of an input structure, as well as hereditarily finite sets built over D, i.e., the set $HF(D)$ of hereditarily finite sets over D is the smallest set such that if $a_1, ..., a_n \in D \cup HF(D)$, then $\{a_1, ..., a_n\} \in D \cup HF(D)$.

In the background of the BGS model, there is a set $\mathcal{D}$ of base domains such that $D = \bigcup \mathcal{D}$. The background signature Υ_K^{BGS} contains a set constructor $\{\cdot\}$ with unfixed arity. Hence, the background class of the BGS model consists of the universe $U^{BGS} = D \cup HF(D)$ and the interpretation of function symbols in Υ_K^{BGS} over U^{BGS}. □

Example 3.1.4. Let us consider the type system used in [3] with some slight modifications. Type expressions are defined by

$$\tau = \tau_\lambda \mid \tau_D \mid \tau_P \mid (A_1 : \tau_1, \ldots, A_k : \tau_k) \mid \{\tau\} \mid \tau_1 \sqcup \tau_2 \mid \tau_1 \sqcap \tau_2$$

.

The interpretation of these type expressions (denoted as $[\![\tau]\!]$) is formally defined as follows:

$$\begin{aligned}
[\![\tau_\lambda]\!] &= \emptyset \\
[\![\tau_D]\!] &= \alpha(\tau_D) \\
[\![\tau_P]\!] &= \beta(\tau_P) \\
[\![(A_1 : \tau_1, ..., A_k : \tau_k)]\!] &= \{(A_1 : a_1, ..., A_k : a_k) \mid a_i \in [\![\tau_i]\!], i = 1, ..., k\} \\
[\![\{\tau\}]\!] &= \{\{a_1, ..., a_j\} \mid j \geq 0 \text{ and } a_i \in [\![\tau]\!], i = 1, ..., j\} \\
[\![\tau_1 \sqcup \tau_2]\!] &= [\![\tau_1]\!] \cup [\![\tau_2]\!] \\
[\![\tau_1 \sqcap \tau_2]\!] &= [\![\tau_1]\!] \cap [\![\tau_2]\!]
\end{aligned}$$

So τ_λ is a trivial type denoting the empty set $\emptyset$. τ_D and τ_P represent a base type for constants and a class type for objects, respectively. α and β are functions mapping each base type to a possibly infinite set of constants, and each class type to a finite set of objects, respectively. There are constructor symbols $(\cdot)$ for finite tuples with bounded arity k, finite sets $\{\cdot\}$ with unfixed arity, as well as unions $\sqcup$ and intersections $\sqcap$, both of arity 2.

In addition to these, the types are associated with function symbols $\in$ of arity 2 denoting set membership, π_i $(1 \leq i \leq k)$ of arity k denoting projection functions on tuples, and $\cup$ and $\cap$, both of arity 2 denoting union and intersection, respectively. □

For every database transformation, a binary tuple constructor $(,)$ is indispensable. This is due to the formalisation of update that is a pair of a location and an update value as will be defined in Definition 3.1.8. Furthermore, the type constructor for finite multisets plays a critical role in the background of a database transformation. A database transformation may have many subcomputations running in parallel, which may yield identical updates. As we will introduce later, by assigning a location operator to a location, a collection of (possibly identical) updates to the location yielded during a computation can be aggregated to a single update to that location. Thus, we need the type constructor for finite multisets to collect all the updates generated by a database transformation in one-step transitions.

The following are several multiset operations from [27]. We use the constructor symbol $\{\!\!\{\cdot\}\!\!\}$ for finite multisets with unfixed arity. Let x and y be two multisets, and $\mathcal{M}$ be a set of multisets. Then,

- $x \uplus y$ returns a multiset that has members from x and y, and the occurrence of each member is the sum of the occurrences of such a member in x and in y.

- $\biguplus \mathcal{M}$ returns a multiset that has members from all elements of $\mathcal{M}$, and the occurrence of each member is the sum of the occurrences of such a member in all elements of $\mathcal{M}$.

- $AsSet(x)$ returns a set that has the same members as x, such that

$$AsSet(x) \quad = \quad \{a|\ a \in x\}$$

- $\mathbf{I}x$ is defined by

$$\mathbf{I}x \quad = \quad \begin{cases} a & \text{if } x = \{\!\!\{a\}\!\!\} \\ \bot & \text{otherwise} \end{cases}$$

Given the base set of a state S, we can add the required Booleans and $\bot$, partition them into base domains, then apply the construction provided in Definition 3.1.7 to obtain a much larger base set, and interpret functions symbols in the signature of S with respect to this enlarged base set.

Postulate 3.1.3 (background postulate). Each state of a database transformation T must contain

- an infinite set of reserve values,
- truth values and their connectives, the equality predicate, the undefinedness value $\bot$, and
- a background class K defined by a background signature Υ_K that contains at least
  - a binary tuple constructor $(,)$, a multiset constructor $\{\!\{\cdot\}\!\}$, and
  - function symbols for operations such as pairing and projection for pairs, and empty multiset $\{\!\{\}\!\}$, singleton $\{\!\{x\}\!\}$, binary multiset union $\uplus$, general multiset union $\biguplus x$, $AsSet$, and $\mathbf{I}x$ ("the unique") on multisets.

The minimum requirements in the background postulate are the same as for parallel algorithms [27], but we do not limit other constructors that will be in the background class in order to capture any request in data models.

3.1.4 Updates

The definitions of location, location content, update, update set and update multiset are the same as for ASMs [47].

Definition 3.1.8. For a database transformation T, let S be a state of T, f be a dynamic function symbol of arity n in the state signature of T and $a_1, ..., a_n, b$ be elements in the base set of S. Then,

- $f(a_1, ..., a_n)$ is called a *location* of T. The interpretation of ℓ in S is called the *content* of ℓ in S, denoted by $val_S(\ell)$.
- An *update* of T is a pair (ℓ, b), where ℓ is a location and b is called the *update value* of ℓ. An *update set* Δ is a set of updates. An *update multiset* $\ddot{\Delta}$ is a multiset of updates.

- An update is *trivial* in a state S if its location content in S is the same as its update value, while an update set is *trivial* if all of its updates are trivial.
- An update set Δ is *consistent* if it does not contain conflicting updates, i.e. for all $(\ell, b), (\ell, b') \in \Delta$ we have $b = b'$.

It is possible to construct for each $(S, S') \in \delta_T$ a minimal update set $\Delta(T, S, S')$ such that applying this update set to the state S will produce the state S'. More precisely, if S is a state of the database transformation T and Δ is a consistent update set for the signature of T, then there exists a unique state $S' = S + \Delta$ resulting from updating S with Δ: we simply have

$$val_{S+\Delta}(\ell) = \begin{cases} b & \text{if } (\ell, b) \in \Delta \\ val_S(\ell) & \text{else} \end{cases}$$

If Δ is not consistent, we let $S + \Delta$ be undefined. Note that this last point is different from the treatment of inconsistent update sets in [82]. However, as discussed there the difference is a mere technicality as long as we concentrate on a single database transformation. The same as the addition of final states in database transformations, the distinction only becomes necessary when placed into the context of persistence with several concurrent database transformations, and a serialisability request. In that case a computation that gets stuck and thus has to be aborted (this is the case, when $S + \Delta$ is undefined) has to be distinguished from a computation that produces the same state S over and over again (this is the case, if $S + \Delta$ is defined as S in case of Δ being inconsistent as in [82]).

Lemma 3.1.1. *Let $S, S' \in \mathcal{S}_T$ be states of the database transformation T with the same base set. Then there exists a unique, minimal consistent update set $\Delta(T, S, S')$ with $S' = S + \Delta(T, S, S')$.*

Note that the minimality of the update set implies the absence of trivial updates.

Proof. Let $Loc_\Delta = \{\ell \mid val_S(\ell) \neq val_{S'}(\ell)\}$ be the set of locations, on which the two states differ. Then the update set $\Delta(T, S, S') = \{(\ell, val_{S'}(\ell) \mid \ell \in Loc_\Delta\}$ is the one needed. □

Let us now look at the one-step transition relation δ_T of a database transformation T. As we permit non-determinism, i.e. there may be more than one successor state of a state S, we need sets of update sets. Therefore, for a database transformation T and a state $S \in \mathcal{S}_T$, we define

$$\Delta(T, S) = \{\Delta(T, S, S') \mid (S, S') \in \delta_T\}$$

.

Let us take a brief look at the effect of isomorphisms on update sets and sets of update sets. For this, any isomorphism ς can be extended to updates $(f(a_1,\dots,a_n),b)$ by defining $\varsigma((f(a_1,\dots,a_n),b)) = (f(\varsigma(a_1),\dots,\varsigma(a_n)),\varsigma(b))$, and to sets by defining $\varsigma(\{u_1,\dots,u_k\}) = \{\varsigma(u_1),\dots,\varsigma(u_k)\}$.

Lemma 3.1.2. *Let S_1 be a state of a database transformation T and ς be an isomorphism from S_1 to S_2. Then $\Delta(T,S_2,\varsigma(S_1')) = \varsigma(\Delta(T,S_1,S_1'))$ for all $(S_1,S_1') \in \delta_T$, and consequently $\Delta(T,S_2) = \varsigma(\Delta(T,S_1))$.*

Proof. According to the abstract state postulate all $(S_2,S_2') \in \delta_T$ have the form $(\varsigma(S_1),\varsigma(S_1'))$ with $(S_1,S_1') \in \delta_T$. Then $val_{\varsigma(S_1)}\varsigma(\ell) = \varsigma(val_{S_1}(\ell))$ and analogously for S_1'. So

$$Loc_{\Delta_2} = \{\ell \mid val_{S_2}(\ell) \neq val_{S_2'}(\ell)\} = \{\varsigma(\ell) \mid val_{S_1}(\ell) \neq val_{S_1'}(\ell)\} = \varsigma(Loc_{\Delta_1}),$$

and

$$\begin{aligned}\Delta(T,S_2,S_2') = &\{(\ell, val_{S_2'}(\ell)) \mid \ell \in Loc_{\Delta_2}\} = \\ &\{(\varsigma(\ell),\varsigma(val_{S_1'}(\ell))) \mid \ell \in Loc_{\Delta_1}\} = \varsigma(\Delta(T,S_1,S_1')).\end{aligned}$$

□

3.1.5 Bounded Exploration

The bounded exploration postulate for sequential algorithms requests that only a finite number of terms can be updated in an elementary step [82]. For parallel algorithms this postulate becomes significantly more complicated, as basic constituents not involving any parallelism (so-called "proclets") have to be considered [27].

For database transformations the problem lies somewhere in between. Computations are intrinsically parallel, even though implementations may be sequential, but the parallelism is restricted in the sense that all branches execute de facto the same computation. We will capture this by means of location operators, which generalise aggregation functions as in [60]. Depending on the data model used and thus on the actual background signature, complex values such as tree-structured values may be involved in computations. As a consequence, we have to cope with the problem of partial updates [89], that is, the synchronisation of updates to different parts of the same tree-structured values or more generally complex database objects.

Location operators define operations on multisets and as such form an important part of logics for meta-finite structures [74]. They allow us to express in a simple way the intrinsic parallelism in many database aggregate functions such as building sums or average values over query results, select maximum or minimum, and even structural recursion on sets, multisets, lists or trees.

Definition 3.1.9. Let D, D' and D'' be the domains, and $\mathcal{M}(D)$ be the set of all non-empty multisets over D. Then a *location operator* ρ over $\mathcal{M}(D)$ consists of a unary function $f_\alpha : D \to D'$, a commutative and associative binary operation $\odot$ over D', and a unary function $f_\beta : D' \to D''$, which define $\rho(b) = f_\beta(f_\alpha(a_1) \odot \cdots \odot f_\alpha(a_n))$ for $b = \{\!\{a_1, ..., a_n\}\!\} \in \mathcal{M}(D)$.

Using a *location function* (denoted by θ) that assigns a location operator or $\perp$ to each location, an update multiset can be reduced to an update set. The idea behind location operators is inspired by the synchronisation of parallel updates in [27]. First, updates generated by parallel computations define an update multiset, then all updates to the same location are merged by means of a location operator to reduce the update multiset to an update set.

Definition 3.1.10. Let $\ddot{\Delta}$ be an update multiset. Then an update set Δ can be obtained by reducing $\ddot{\Delta}$ such that

$$\begin{aligned} \Delta = \{(\ell, b) | \theta(\ell) = \rho \text{ and } b = \rho(\biguplus_{(\ell, b') \in \ddot{\Delta}} \{\!\{b'\}\!\})\} \\ \cup \{(\ell, b) | \theta(\ell) = \perp \text{ and } (\ell, b) \in \ddot{\Delta}\} \end{aligned}$$

We provide several examples to illustrate that location operators can be applied in many different ways. They turn out to be a powerful tool to handle dynamically-constructed values in parallel.

Example 3.1.5. The standard aggregate functions provided in relational databases are location operators. Let us have a look at *sum* and *avg* in detail.

- *sum* is a location operator using the unary function $f_\alpha(a) = a$, the commutative and associative $+$-operation for $\odot$, i.e., $a_1 \odot a_2 = a_1 + a_2$, and the unary function $f_\beta(a) = a$.
- *avg* is also a location operator with $f_\alpha(a) = (a, 1)$, $(a_1, b_1) \odot (a_2, b_2) = (a_1 + a_2, b_1 + b_2)$, and $f_\beta(a, b) = a \div b$.

□

Example 3.1.6. *XMLAgg* is an aggregate function supported by Oracle XML DB, which produces a forest of XML elements from a collection of XML elements. The following statement produces a Publications element containing Title elements whose contents are the titles of publications in the academic unit "Computer Science". Moreover, the titles of publications are ordered by their publication IDs.

```
SELECT XMLElement("Publications",
         XMLAgg(XMLElement("Title",a.Title)
                ORDER BY a.PubID))
    AS ComputerScienceReport
  FROM Publication a,
       AcademicUnit b,
       Authorship c
 WHERE c.UnitID = b.UnitID
   AND a.PubID=c.PubID
   AND b.Name = "Computer Science";
```

The result of executing the above statement can be something as follows:

ComputerScienceReport

```
------------------------------------------------------------
<Publications>
  <Title>XML Machines</Title>
  <Title>A Hedge Algebra for XML Trees</Title>
  <Title>Abstract State Services</Title>
  <Title>Partial Updates in Complex-value Databases</Title>
</Publications>
```

XMLAgg is a location operator with $f_\alpha(a) = a$ (i.e., the identity function), $\odot$ as the operation for returning a forest of XML elements along with the values used for ordering them if the ORDER BY clause is used, i.e., $(a_1, b_1) \odot (a_2, b_2) = \{\!\{(a_1, b_1), (a_2, b_2)\}\!\}$, and $f_\beta(a) = a'$ for $a = \{\!\{(a_1, b_1), ..., (a_n, b_n)\}\!\}$ generating a forest of XML elements referring to $a_1, ..., a_n$ in an order given by $b_1, ..., b_n$ in a.

The nested calls to *XMLAgg* can be used to reflect the hierarchical structure among elements. When using *XMLElement* and *XMLAgg* together, all of the XML fragments identified by the query can be aggregated into a single, well-formed XML document. □

Example 3.1.7. Consider the evaluation of a first-order formula $\forall x \in D_1 \exists y \in D_2 \varphi(x, y)$. Assume that the evaluation result will be stored at the location ℓ. Let the cardinalities of D_1 and D_2 be n_1 and n_2, respectively. Then there are two nested parallel computations involved in the evaluation. The inner parallel computation for a specific value $a_i \in D_1$ ($i \in [1, n_1]$) has n_2 parallel branches, each of which evaluates a term $\varphi(a_i, b_j)$ for the various values $b_j \in D_2$ ($j \in [1, n_2]$), thus producing an update multiset with n_2 updates $(\ell, true)$ or $(\ell, false)$. Using $\theta(\ell) = \bigvee$ (logical OR) to assign a location operator $\bigvee$ to ℓ – in this case f_α and f_β are the identity function, and $\odot$ is $\vee$ – evaluates the inner existentially quantified formula, thereby producing another update multiset with n_1 entries $(\ell, true)$ or $(\ell, false)$. Using $\theta(\ell) = \bigwedge$ (logical AND) to assign a location operator $\bigwedge$ to ℓ again, this update multiset is reduced to a set with a single update. □

If a database transformation T uses location operators, they must be defined in the algorithmic part of the state signature of T.

The problem of partial updates [87, 88, 89] refers to computations over complex-value databases that may produce updates at different levels of abstraction. An immediate consequence of such updates is that atomicity of locations is no longer guaranteed in general, and clashes may arise due to the overlapping, subsumption or dependence among locations. Chapter 6 will be dedicated to the discussion of the partial update problem in detail. However, the issues relating to inconsistent update sets are irrelevant to the proof of the characterisation theorem in Section 3.3 as pointed out in [27, 82]. In fact, the problem of partial updates is subsumed by the problem of providing consistent update sets, in which there cannot be pairs (ℓ, b_1) and (ℓ, b_2) with $b_1 \neq b_2$.

The bounded exploration postulate in [82] for sequential algorithms is motivated by the *sequential accessibility principle*, which could be phrased as the request that each location must be uniquely identifiable. Leaving aside the discussion of how to deal logically with partially defined terms, unique identifiability can be obtained by using terms of the form $\mathrm{I}x.\varphi(x)$ with a formula φ, in which x is the only free variable. Such terms have to be interpreted as "the unique x satisfying formula $\varphi(x)$", which of course may be undefined if no such x exists or more than one exist. According to the abstract state postulate defined for database transformations the sequential accessibility principle must be preserved for the algorithmic part of a meta-finite state.

In principle, the claim of unique identifiability also applies to databases, as emphasised by Beeri and Thalheim in [21]. Unique identifiability has to be claimed for the basic updatable units in a database, for example, objects in [124]. Unique identifiability, however, does not necessarily apply to all elements in a database. Sets of logically indistinguishable locations may be updated simultaneously. Nevertheless, for databases only logical properties are relevant – this is the so-called "genericity principle" in database theory [20] – and therefore, it must still be possible to use terms to access elements and locations in the database part of a state. These terms, however, may be non-ground, containing variables. If a non-ground term identifies more than one location in a state S, these locations will be called accessible in parallel.

Definition 3.1.11. Let S be a state of the database transformation T.

- An element a of S is *accessible* if there is a ground term t in the signature of S that is interpreted as a in the state S.
- A location $f(a_1, \dots, a_n)$ is *accessible* if the elements $a_1, \dots, a_n$ are all accessible.
- An update $(f(a_1, \dots, a_n), b)$ is *accessible* if the location $f(a_1, \dots, a_n)$ and the element b are accessible.
- Locations $f(a_1^1, \dots, a_n^1), \dots, f(a_1^m, \dots, a_n^m)$ with $f \in \Upsilon_{db}$ are *accessible in parallel* if there exists a term t' and an accessible element b', such that the values for which t' is interpreted by b' in S are $f(a_1^1, \dots, a_n^1), \dots, f(a_1^m, \dots, a_n^m)$.
- Updates $(f(a_1^1, \dots, a_n^1), b), \dots, (f(a_1^m, \dots, a_n^m), b)$ with $f \in \Upsilon_{db}$ are *accessible in parallel* if $f(a_1^1, \dots, a_n^1), \dots, f(a_1^m, \dots, a_n^m)$ are accessible in parallel and b is accessible.

The first three items of Definition 3.1.11 are exactly the same as defined in [82, Definition 5.3]. The last two items formalise our discussion above.

Example 3.1.8. Take a database transformation T with a ternary predicate symbol R in its signature. Let the interpretation of R in a state S be $\{(a, a, b), (a, b, c), (b, b, a), (b, a, c)\}$. Then $R(a, a, b)$ and $R(b, b, a)$ are accessible in parallel using the term $R(x, x, y)$ and the accessible element **true**. □

The bounded exploration postulate in the sequential ASM thesis [82] uses a finite set of ground terms as a *bounded exploration witness* in the sense that whenever states S_1 and S_2 coincide over this set of ground terms the update set produced by the sequential algorithm is the same in these states. The intuition behind the postulate is that only the part of a state that is given by means of the bounded exploration witness will actually be explored by the algorithm.

The fact that only finitely many locations can be explored remains the same for database transformations. However, permitting parallel accessibility within the database part of a state forces us to slightly change our view on the bounded exploration witness. For this we need access terms.

A variable assignment ζ for state S is a finite function which assigns elements in the base set of S to a finite number of variables. If ζ is a variable assignment, then $\zeta[x_1 \mapsto b_1, \ldots, x_k \mapsto b_k]$ is another variable assignment defined by

$$\zeta[x_1 \mapsto b_1, \ldots, x_k \mapsto b_k](x) = \begin{cases} b_i & \text{if } x = x_i (i = 1, \ldots, k) \\ \zeta(x) & \text{else} \end{cases}$$

For convenience, we use the notation $val_{S,\zeta}(t)$ for the interpretation of a term t in a state S under a variable assignment ζ, and refer to *database variables* as variables that must be interpreted by elements in B_{db}.

Definition 3.1.12. An *access term* is either a ground term t_α or a pair (t_β, t_α) of terms, the variables $x_1, \ldots, x_n$ in which must be database variables, referring to the arguments of some dynamic function symbol $f \in \Upsilon_{db} \cup \{f_1, ..., f_\ell\}$.

- The interpretation of (t_β, t_α) in a state S is the set of locations

$$\{f(a_1, \ldots, a_n) \mid val_{S,\zeta}(t_\beta) = val_{S,\zeta}(t_\alpha) \text{ with } \zeta = \{x_1 \mapsto a_1, \ldots, x_n \mapsto a_n\}\}.$$

Structures S_1 and S_2 *coincide* over a set $\mathcal{T}_{wit}$ of access terms if the interpretation of each $(t_\beta, t_\alpha) \in \mathcal{T}_{wit}$ over S_1 and S_2 are equal.

Instead of writing (t_β, t_α) for an access term, we should in fact write (f, t_β, t_α). For simplicity we drop the function symbol f and assume it is implicitly given.

Due to our request that the database part of a state is always finite there will be a maximum number n_1 of elements that are accessible in parallel. Furthermore, there is always a number n_2 such that n_2 variables are sufficient to describe the updates of a database transformation, and n_2 can be taken to be minimal. Then for each state S the upper boundary of exploration is $\mathcal{O}(n_1^{n_2})$, where n_1 depends on S. Taking these together we obtain our fourth postulate.

Postulate 3.1.4 (bounded exploration postulate). For a database transformation T there exists a fixed, finite set $\mathcal{T}_{wit}$ of access terms of T such that $\Delta(T, S_1) = \Delta(T, S_2)$ holds whenever the states S_1 and S_2 coincide over $\mathcal{T}_{wit}$.

Same as in the sequential ASM thesis we continue calling the set $\mathcal{T}_{wit}$ of access terms a *bounded exploration witness*. The only difference to the bounded exploration postulate for sequential algorithms in [82] is the use of access terms (t_β, t_α), whereas in the sequential ASM thesis only ground terms are considered. Access terms of the form (t_β, t_α) are actually equivalent to closed set comprehension terms $\{f(x_1, \ldots, x_n) \mid t_\beta = t_\alpha\}$, i.e. they express first-order queries similar to relational calculus. Due to the fact that the database part of a state is a finite structure the set of locations defined by an access term is always finite. However, building terms on top of the state signature does not yet capture such terms. Access terms for the algorithmic part can still only be ground terms, otherwise finiteness cannot be guaranteed. Therefore, the modified Postulate 3.1.4 still expresses the same intention that the bounded exploration postulate for sequential algorithms does, i.e. only finitely many locations can be updated at a time, and these locations are determined by finitely many terms that appear in some way in the textual description of the database transformation.

3.1.6 Bounded Non-determinism

The last postulate addresses the question of how non-determinism is permitted in a database transformation. To handle this, we need to further clarify the relationship between access terms and the internal structure of states. As defined in the abstract state postulate, a state of a database transformation is a meta-finite structure consisting of two parts: the database part and algorithmic part, which are linked via a fixed, finite number of bridge functions. Hence, in terms of a database transformation T with a set of its states, we consider that a *ground access term* t_α of T is allowed to only access the algorithmic part of states, while a *non-ground access term* (t_β, t_α) may access either the database or the algorithmic parts.

Example 3.1.9. Let us consider Example 3.1.1, in which we have $\text{PERSON} \in \Upsilon_{db}$, EVEN and $+ \in \Upsilon_a$, f_{num} as a bridge function and $B_a = \mathbb{R} \cup \Sigma$. The terms 2, 7.8, $+(3, 9)$ and $\text{EVEN}(9)$ may become ground access terms. The terms $(\text{PERSON}(x, y, z, z'), \textbf{true})$, $(f_{num}(x), 8)$, $(+(f_{num}(x), f_{num}(y)), 20)$ and $(\text{EVEN}(f_{num}(x)), \textbf{false})$ may become non-ground access terms.

We formally elaborate these two kinds of access terms in terms of a signature $\Upsilon = \Upsilon_{db} \cup \Upsilon_a \cup \{f_1, ..., f_\ell\}$ and a base set B of states, i.e., $B = B_{db} \cup B_a$. A *ground access term* t_α is a ground term that can be defined as follows:

- $t \in B_a$ is a ground term, and
- $f(t_1, ..., t_n)$, for n-ary function symbol $f \in \Upsilon_a$ and ground terms $t_1, ..., t_n$, is a ground term.

A *non-ground access term* (t_β, t_α) is a pair of terms, in which at least one of them is a non-ground term that is inductively defined by applying function symbols from Υ over database variables in accordance with the definition of a meta-finite structure as in Definition 3.1.6.

Before explaining how the bound of non-determinism is imposed in database transformations, we define equivalent substructures in the following sense.

Definition 3.1.13. Given two structures S' and S of the same signature Υ, a structure S' is a *substructure* of the structure S (notation: $S' \preceq S$) if

- the base set B' of S' is a subset of the base set B of S, i.e., $B' \subseteq B$, and
- for each function symbol f of arity n in the signature Υ the restriction of $val_S(f(x_1, ..., x_n))$ to B' results in $val_{S'}(f(x_1, ..., x_n))$.

Substructures $S_1, S_2 \preceq S$ are *equivalent* (notation: $S_1 \equiv S_2$) if there exists an automorphism $\varsigma \in Aut(S)$ with $\varsigma(S_1) = S_2$. The *equivalence class* of a substructure S' in the structure S is the subset of all substructures of S which are equivalent to S'.

Example 3.1.10. Let us consider a simple ternary relation schema R. Suppose our database contains $R(a, a, b_1), R(b, b, a_1), R(c_1, c, c_2)$. Then $R(a, a, b_1)$ defines a substructure with base set $\{a, b_1\}$. This substructure is equivalent to the substructure $R(b, b, a_1)$, as the isomorphism defined by the permutation $(a, b)(b_1, a_1)$ just swaps the two substructures.

If, however, we have a second relation schema R', and the database contains only $R'(a, b_1)$ and $R'(c, c)$, then the restriction to $\{a, b_1\}$ defines a substructure containing $R(a, a, b_1)$ and $R'(a, b_1)$, whereas the restriction to $\{b, a_1\}$ defines a substructure $R(b, b, a_1)$ – these substructures are no longer equivalent.

If the database contained also $R'(b, a_1)$, the two restrictions would again define equivalent substructures. □

Let us start with the case where states have no bridge functions. We want to restrict non-determinism to only accessing elements in the database part of a state that is a finite structure. By the abstract state postulate, different branches reflect all the possible choices, which thus captures the genericity of database transformations (i.e., preserved under isomorphisms [20]). Any isomorphism between states gives rise to an isomorphism between successor states. Consequently, whenever a substructure of the database part is preserved in some successor state, then each equivalent substructure of the database part is preserved in some (possibly same) successor state. This is because the automorphism that interchanges two equivalent substructures permutes successor states, according to Definition 3.1.13. Nevertheless, it is also possible that there is a successor state, in which none of the equivalent substructures in the database part is preserved.

Example 3.1.11. Let us consider the relation $R = \{(a, a, b_1), (b, b, a_1), (c_1, c, c_2)\}$ in Example 3.1.10 again.

- Suppose that we non-deterministically delete a tuple from R in a state S. Then there will be three successor states of S with $R = \{(b, b, a_1), (c_1, c, c_2)\}$, $R = \{(a, a, b_1), (c_1, c, c_2)\}$ or $R = \{(a, a, b_1), (b, b, a_1)\}$, respectively. It is clear to see that not all of the successor states are isomorphic.
- If we non-deterministically select two tuples from R in a state S and delete them. Then we will also get three successor states of S: $R = \{(b, b, a_1)\}$, $R = \{(a, a, b_1)\}$ or $R = \{(c_1, c, c_2)\}$. In this case, in terms of the equivalence class $\{(a, a, b_1), (b, b, a_1)\}$, none of the equivalent substructures are preserved in the successor state of S with $R = \{(c_1, c, c_2)\}$.

In the case that states have bridge functions, however, the situation becomes a bit tricky because bridge functions define substructures of the algorithmic part based on substructures of the database part. Thus non-determinism caused by non-deterministically selecting elements in the database part may also result in the non-deterministic changes on substructures of the algorithmic part. Nevertheless, the distinction between the database and algorithmic parts is that non-determinism cannot arise from the algorithmic part by selecting non-deterministically substructures to the algorithmic part of a state.

Example 3.1.12. Let us consider the relation $R = \{(a, a, b_1), (b, b, a_1), (c_1, c, c_2)\}$ in Example 3.1.10. Assume that there exist a bridge function $f_{num} = \{(a, 4), (a_1, 6), (b, 3), (b_1, 1)(c, 5), (c_1, 8), (c_2, 7)\}$ and $\{\text{Even}, \text{Odd}, \text{Test}\} \subseteq \Upsilon_a$.

We may use the non-ground access terms $(R(x, y, z), \mathbf{true})$, $(\textsc{Even}(f_{num}(x)), \mathbf{true})$ and $(\textsc{Odd}(f_{num}(z)), \mathbf{true})$ with the formula $R(x, y, z) \wedge \textsc{Even}(f_{num}(x)) \wedge \textsc{Odd}(f_{num}(z))$ to retrieve out the tuples (a, a, b_1) and (c_1, c, c_2) in R.

Then we can non-deterministically generate updates on function $\textsc{Test}$ by non-deterministically selecting one of these two tuples. For example, two update sets $\{(\textsc{Test}(4), 1)\}$ and $\{(\textsc{Test}(8), 7)\}$ may be created by using $(\textsc{Test}(x), z)$ together with the formula $R(x, y, z) \wedge \textsc{Even}(f_{num}(x)) \wedge \textsc{Odd}(f_{num}(z))$.

Therefore, substructures of its algorithmic part may or may not be preserved in the successor states of a state. This indeed can be explained under a border view on equivalence classes defined in terms of a state taking all of the database part, the algorithmic part and bridge functions into consideration. The rationale that whenever a substructure of a state is preserved in some successor state, each equivalent substructure of a state is preserved in some (possibly same) successor state is still captured by the abstract state postulate in the same way as in the case without bridge functions.

Now we formalise the bounded non-determinism postulate to capture these ideas by properly defining the presence of non-ground access terms. In doing so, we put a severe restriction on the non-determinism in the transition relation δ_T.

Postulate 3.1.5 (bounded non-determinism postulate). If there are states S_1, S_2 and $S_3 \in \mathcal{S}_T$ with $(S_1, S_2) \in \delta_T$, $(S_1, S_3) \in \delta_T$ and $S_2 \neq S_3$, then there exists an access term of the form (t_β, t_α) in $\mathcal{T}_{wit}$.

According to this postulate, if a database transformation T over some state S_1 has non-determinism (i.e., $\Delta(T, S_1)$ contains more than one update set), then we must have a non-ground access term in the bounded exploration witness of T. Alternatively, if the bounded exploration witness of T contains only ground access terms, then T can only access the algorithmic part of a state and cannot have non-determinism. The bounded non-determinism postulate is motivated by the intrinsic non-determinism in database queries and updates that permit identifier creation. This will become clear in the proof of our main result in Section 3.3, according to which the bounded non-determinism postulate enforces that only the bounded choice among database elements can be the source of non-determinism.

Remark 3.1.1. Van den Bussche defined the notions of *determinacy* and *semi-determinism*: a determinate query preserves the input database, whereas a semi-deterministic database transformation produces isomorphic outputs and thus preserves the input database up to an automorphism [146, 148]. In the sense of Van den Bussche an input database would define a substructure, and by the bounded non-determinism postulate preserving this substructure implies that each automorphism of the input database defines an isomorphism between the possible successor states, but not all of successor states will be isomorphic. Hence, database transformations characterised by five postulates subsume semi-deterministic transformations. In the same way they captures the insertion of new objects with a choice of identifiers as worked out for generic updates in [124].

3.1.7 Final Remarks

Naturally, for a database transformation, the decisive part is the progression of the database part of states, whereas the algorithmic part could be understood as playing only a supporting role. Nonetheless, the postulates for database transformations permit transformations, in which the major computation happens on the algorithmic part. In the extreme case we could even only manipulate the algorithmic part. This implies that our model actually subsumes all sequential algorithms. Furthermore, all extensions such as genericity, meta-finite states, location operators, bounded exploration with non-ground terms and bounded non-determinism only affect the database part. This will become more apparent in the next two sections, when we present a variant of ASMs capturing exactly database transformations as stipulated by the five postulates. On the other hand, our model of database transformations does not capture parallel algorithms, as the bounded exploration postulate excludes unbounded parallelism.

3.2 Database Abstract State Machines

In this section we define a variant of Abstract State Machines, which we call *database Abstract State Machines*, and show that DB-ASMs satisfy the postulates of a database transformation. In the next section we will address the more challenging problem showing the converse of this result.

3.2.1 DB-ASM Rules and Update Sets Generated by Them

We first define DB-ASM rules. If r is a DB-ASM rule and S is a state (i.e., a Υ-structure for the signature Υ of r), we associate a set $\Delta(r, S)$ of update sets with r and S. For convenience, we also use the notation $\ddot{\Delta}(r, S)$ for a set of update multisets defined by r

and S. DB-ASM rules may involve variables, so in the following definition we also use the notation $\Delta(r, S, \zeta)$ for a set of update sets that depends on a variable assignment ζ, and analogously $\ddot{\Delta}(r, S, \zeta)$ for a set of update multisets.

For the signature Υ, we adopt the requirements of the abstract state postulate. That is, it comprises a sub-signature Υ_{db} for the database part, a sub-signature Υ_a for the algorithmic part, and bridge functions $\{f_1, \dots, f_\ell\}$. For states we assume that the requirement in the abstract state postulate, according to which the restriction to Υ_{db} results in a finite structure, is satisfied. Furthermore, we assume a background in the sense of the background postulate being defined.

The notation $var(t)$ is used to denote the set of variables occurring in a term t. Similar to free variables occurring in formulae we can define the set $fr(r)$ of free variables appearing in a DB-ASM rule r, with variables being bound by forall and choice rules. A rule r is called *closed* iff $fr(r) = \emptyset$.

Definition 3.2.1. The set $\mathcal{R}$ of *DB-ASM rules* over a signature $\Upsilon = \Upsilon_{db} \cup \Upsilon_a \cup \{f_1, \dots, f_\ell\}$ and associated sets of update sets (with respect to states as in Postulate 3.1.2 with a background as in Postulate 3.1.3) are defined as follows:

- If $t_0, \dots, t_n$ are terms over Υ, and f is a n-ary dynamic function symbol in Υ, then $f(t_1, \dots, t_n) := t_0$ is a rule r in $\mathcal{R}$ called *assignment rule* with $fr(r) = \bigcup_{i=0}^{n} var(t_i)$, where $var(t_i)$ is the set of variables occurring in the terms t_i $(i = 0, \dots, n)$.

  For a state S over Υ and a variable assignment ζ for $fr(r)$ we obtain

  $$\Delta(r, S, \zeta) = \{\{(f(a_1, \dots, a_n), a_0)\}\}$$

  with $a_i = val_{S,\zeta}(t_i)$ $(i = 0, \dots, n)$, and

  $$\ddot{\Delta}(r, S, \zeta) = \{\{\!\{(f(a_1, \dots, a_n), a_0)\}\!\}\}$$

- If φ is a first-order formula of Υ without free variables and $r' \in \mathcal{R}$ is a DB-ASM rule, then **if** φ **then** r' **endif** is a rule r in $\mathcal{R}$ called *conditional rule* with $fr(r) = fr(\varphi) \cup fr(r')$.

  For a state S over Υ and a variable assignment ζ for the variables in $fr(r)$, we obtain

  $$\ddot{\Delta}(r, S, \zeta) = \begin{cases} \ddot{\Delta}(r', S, \zeta) & \text{if } val_{S,\zeta}(\varphi) = true \\ \emptyset & \text{else} \end{cases}$$

and

$$\Delta(r,S,\zeta) = \begin{cases} \Delta(r',S,\zeta) & \text{if } val_{S,\zeta}(\varphi) = true \\ \emptyset & \text{else} \end{cases}$$

- If φ is a first-order formula of Υ with only database variables, $\{x_1,\dots,x_k\} \subseteq fr(\varphi)$ and $r' \in \mathcal{R}$ is a DB-ASM rule, then **forall** $x_1,\dots,x_k$ **with** φ **do** r' **enddo** is a rule r in $\mathcal{R}$ called *forall rule* with $fr(r) = (fr(r') \cup fr(\varphi)) - \{x_1,\dots,x_k\}$.

  For a state S over Υ and a variable assignment ζ for the variables in $fr(r)$, let $\mathcal{B} = \{(b_1,\dots,b_k) \mid val_{S,\zeta[x_1\mapsto b_1,\dots,x_k\mapsto b_k]}(\varphi) = true\}$ and $\mathfrak{M}$ denote the set of mappings α from $\mathcal{B}$ to $\bigcup\{\Delta(r',S,\zeta[x_1 \mapsto b_1,\dots,x_k \mapsto b_k]) \mid (b_1,\dots,b_k) \in \mathcal{B}\}$ with $\alpha(b_1,\dots,b_k) \in \Delta(r',S,\zeta[x_1 \mapsto b_1,\dots,x_k \mapsto b_k])$. Then each $\alpha \in \mathfrak{M}$ defines an update set $\Delta_\alpha = \bigcup\{\alpha(b_1,\dots,b_k) \mid (b_1,\dots,b_k) \in \mathcal{B}\}$, from which we obtain

$$\Delta(r,S,\zeta) = \{\Delta_\alpha \mid \alpha \in \mathfrak{M}\}.$$

  Analogously, let $\ddot{\mathfrak{M}}$ denote the set of mappings $\ddot{\alpha}$ from $\mathcal{B}$ to $\bigcup\{\ddot{\Delta}(r',S,\zeta[x_1 \mapsto b_1,\dots,x_k \mapsto b_k]) \mid (b_1,\dots,b_k) \in \mathcal{B}\}$ with $\ddot{\alpha}(b_1,\dots,b_k) \in \ddot{\Delta}(r',S,\zeta[x_1 \mapsto b_1,\dots,x_k \mapsto b_k])$. Then each $\ddot{\alpha} \in \ddot{\mathfrak{M}}$ defines an update multiset $\ddot{\Delta}_{\ddot{\alpha}} = \biguplus\{\ddot{\alpha}(b_1,\dots,b_k) \mid (b_1,\dots,b_k) \in \mathcal{B}\}$, which finally gives

$$\ddot{\Delta}(r,S,\zeta) = \{\ddot{\Delta}_{\ddot{\alpha}} \mid \ddot{\alpha} \in \ddot{\mathfrak{M}}\}.$$

- If $r_1,\dots,r_n$ are rules in $\mathcal{R}$, then the rule r defined as **par** $r_1 \dots r_n$ **endpar** is a rule in $\mathcal{R}$, called *parallel rule* with $fr(r) = \bigcup_{i=1}^{n} fr(r_i)$.

  For a state S over Υ and a variable assignment ζ for the variables in $fr(r)$ we obtain

$$\Delta(r,S,\zeta) = \{\Delta_1 \cup \dots \cup \Delta_n \mid \Delta_i \in \Delta(r_i,S,\zeta) \text{ for } i = 1,\dots,n\}$$

  and

$$\ddot{\Delta}(r,S,\zeta) = \{\ddot{\Delta}_1 \uplus \dots \uplus \ddot{\Delta}_n \mid \ddot{\Delta}_i \in \ddot{\Delta}(r_i,S,\zeta) \text{ for } i = 1,\dots,n\}.$$

- If φ is a first-order formula of Υ with only database variables, $\{x_1,\dots,x_k\} \subseteq fr(\varphi)$ and $r' \in \mathcal{R}$ is a DB-ASM rule, then **choose** $x_1,\dots,x_k$ **with** φ **do** r' **enddo** is a rule r in $\mathcal{R}$ called *choice rule* with $fr(r) = (fr(r') \cup fr(\varphi)) - \{x_1,\dots,x_k\}$.

For a state S over Υ and a variable assignment ζ for the variables in $fr(r)$ let $\mathcal{B} = \{(b_1, \dots, b_k) \mid val_{S,\zeta[x_1 \mapsto b_1, \dots, x_k \mapsto b_k]}(\varphi) = true\}$. Then we obtain

$$\Delta(r, S, \zeta) = \bigcup \{\Delta(r', S, \zeta[x_1 \mapsto b_1, \dots, x_k \mapsto b_k]) \mid (b_1, \dots, b_k) \in \mathcal{B}\}.$$

and

$$\ddot{\Delta}(r, S, \zeta) = \bigcup \{\ddot{\Delta}(r', S, \zeta[x_1 \mapsto b_1, \dots, x_k \mapsto b_k]) \mid (b_1, \dots, b_k) \in \mathcal{B}\}.$$

- If r_1, r_2 are rules in $\mathcal{R}$, then the rule r defined as **seq** r_1 r_2 **endseq** is a rule in $\mathcal{R}$, called *sequence rule* with $fr(r) = fr(r_1) \cup fr(r_2)$.

  For a state S over Υ and a variable assignment ζ for the variables in $fr(r)$, we obtain

  $$\Delta(r, S, \zeta) = \{\Delta_1 \oslash \Delta_2 \mid \Delta_1 \in \Delta(r_1, S, \zeta) \text{ and } \Delta_2 \in \Delta(r_2, S + \Delta_1, \zeta)\}$$

  with update sets defined as

  $$\Delta_1 \oslash \Delta_2 = \Delta_2 \cup \{(\ell, v) \in \Delta_1 \mid \neg \exists v'.(\ell, v') \in \Delta_2\},$$

  and

  $$\ddot{\Delta}(r, S, \zeta) = \{\ddot{\Delta}_1 \oslash \ddot{\Delta}_2 \mid \ddot{\Delta}_1 \in \ddot{\Delta}(r_1, S, \zeta) \text{ and } \ddot{\Delta}_2 \in \ddot{\Delta}(r_2, S + AsSet(\ddot{\Delta}_1), \zeta)\}$$

  with update multisets defined as

  $$\ddot{\Delta}_1 \oslash \ddot{\Delta}_2 = \ddot{\Delta}_2 \uplus \{\!\{(\ell, v) \in \ddot{\Delta}_1 \mid \neg \exists v'.(\ell, v') \in \ddot{\Delta}_2\}\!\}.$$

- If $r' \in \mathcal{R}$ is a DB-ASM rule and θ is a location function that assigns location operators ρ to terms t with $var(t) \subseteq fr(r')$, then **let** $\theta(t) = \rho$ **in** r' **endlet** is a DB-ASM rule $r \in \mathcal{R}$ called *let rule* with $fr(r) = fr(r')$.

  For a state S over Υ and a variable assignment ζ for the variables in $fr(r)$ let $\ddot{\Delta}(r', S, \zeta) = \{\ddot{\Delta}_1, \dots, \ddot{\Delta}_n\}$ with update multisets $\ddot{\Delta}_i = \ddot{\Delta}_i^{(t)} \uplus \ddot{\Delta}_i^{-}$ such that the first of these two multisubsets contains the updates to locations $val_{S,\zeta}(t)$, while the second one contains updates to all other locations. We define

  $$\ddot{\Delta}_i^{(r)} = \{(\ell, a) \mid \ell = val_{S,\zeta}(t), a = \theta(t)(\{\!\{a_1, \dots, a_k \mid (\ell, a_i) \in \ddot{\Delta}_i^{(t)}\}\!\})\} \uplus \ddot{\Delta}_i^{-}$$

and

$$\Delta_i^{(r)} = \{(\ell, a) \mid \ell = val_{S,\zeta}(t), a = \theta(t)(\{\!\{a_1, \ldots, a_k \mid (\ell, a_i) \in \ddot{\Delta}_i^{(t)}\}\!\})\} \cup \Delta_i^-,$$

with $\Delta_i^- = \{(\ell, a) \mid (\ell, a) \in \ddot{\Delta}_i^-\}$. This finally gives

$$\ddot{\Delta}(r, S, \zeta) = \{\ddot{\Delta}_1^{(r)}, \ldots, \ddot{\Delta}_n^{(r)}\} \quad \text{and} \quad \Delta(r, S, \zeta) = \{\Delta_1^{(r)}, \ldots, \Delta_n^{(r)}\}.$$

Note that only assignment rules "create" updates in update sets and multisets, only choice rules introduce non-determinism, let rules reduce update multisets to update sets by letting updates to the same location collapse to a single update using assigned location operators, whereas all other rules just rearrange these updates into different sets and multisets, respectively. The sequence operator **seq** is associative, so we can also use more complex sequence rules **seq** $r_1 \ldots r_n$ **endseq**.

Example 3.2.1. Consider the following DB-ASM rule

> **forall** x **with** $\exists z.R(x, x, z)$
> **do**
>   **let** $\theta(f(x)) =$ `sum` **in**
>   **forall** y **with** $R(x, x, y)$
>   **do**
> $f(x) := 1$
>   **enddo**
>   **endlet**
> **enddo**

using `sum` as a shortcut for the location operator $(id, +, id)$. If the state contains the tuples $R(a, a, b), R(a, a, b'), R(c, c, a'), R(b, b, c), R(b, b, a'), R(b, b, b)$, $R(b', a', a)$, then first the update multisets $\{\!\{(a, 1), (a, 1)\}\!\}, \{\!\{(c, 1)\}\!\}, \{\!\{(b, 1), (b, 1), (b, 1)\}\!\}$ are produced by means of the forall rules, which are then collapsed to the update set $\{(a, 2), (c, 1), (b, 3)\}$ using the `sum`-operator in the let rule. Thus, for x such that there are tuples $R(x, x, y)$ in the database, then number of such tuples is counted and assigned to $f(x)$.

Let us denote the inner and outer forall rules as r_1 and r_2, respectively. Then we have $fr(r_1) = (\{x\} \cup fr(R(x, x, y))) - \{y\} = \{x\}$ and $fr(r_2) = (\{x\} \cup fr(\exists z.R(x, x, z))) - \{x\} = \emptyset$. Hence this DB-ASM rule is closed. □

Lemma 3.2.1. *Let r be a DB-ASM rule and $\varsigma : S_1 \to S_2$ be an isomorphism between states S_1 and S_2. Let $S_1' = S_1 + \Delta$ be a successor state of S_1 for some $\Delta \in \Delta(r, S_1)$. Then* *we have* $\varsigma(\Delta) \in \Delta(r, S_2)$, *and* $\varsigma : S_1' \to S_2' = S_2 + \varsigma(\Delta)$ *is an isomorphism between the successor states* S_1' *and* S_2'.

Proof. We proceed by structural induction on the rule r. So we start with an assignment rule $f(t_1, \dots, t_n) := t_0$. Then we must take $\Delta = \{(f(a_1, \dots, a_n), a_0)\}$ with $a_i = val_{S_1}(t_i)$ for $i = 0, \dots, n$. For any location ℓ, we have

$$val_{S_1'}(\ell) = \begin{cases} a_0 & \text{if } \ell = f(a_1, \dots, a_n) \\ val_{S_1}(\ell) & \text{else} \end{cases}$$

.

With $\varsigma(\Delta) = \{(f(\varsigma(a_1), \dots, \varsigma(a_n)), \varsigma(a_0))\}$ we obtain $val_{S_2'}(\varsigma(\ell)) = \begin{cases} \varsigma(a_0) & \text{if } \ell = f(a_1, \dots, a_n) \\ val_{S_2}(\varsigma(\ell)) & \text{else} \end{cases} = \varsigma(val_{S_1'}(\ell))$, which gives $S_2' = \varsigma(S_1')$ as desired. The same argument applies to update multisets.

For a conditional rule $r = \textbf{if } \varphi \textbf{ then } r' \textbf{ endif}$, let $\zeta_2 = \varsigma \circ \zeta_1$. Then $val_{S_1,\zeta_1}(\varphi) = true$ iff $val_{S_2,\zeta_2}(\varphi) = true$. This implies

$$\begin{aligned} & S_2 + \Delta(r, S_2, \zeta_2) \\ = & \begin{cases} S_2 + \varsigma(\Delta(r', S_1, \zeta_1)) & \text{if } val_{S_2,\zeta_2}(\varphi) = true \\ S_2 & \text{else} \end{cases} \\ = & \begin{cases} \varsigma(S_1 + \Delta(r', S_1, \zeta_1)) & \text{if } val_{S_1,\zeta_1}(\varphi) = true \\ \varsigma(S_2) & \text{else} \end{cases} \\ = & \ \varsigma(S_1 + \Delta(r, S_1, \zeta_1)). \end{aligned}$$

The other cases are proven analogously.

□

3.2.2 Database Abstract State Machines

We are now prepared to define DB-ASMs and show that they satisfy the five postulates for database transformations from the previous section.

Definition 3.2.2. A *database Abstract State Machine* (DB-ASM) M over signature Υ of states as in Postulate 3.1.2 and with a background as in Postulate 3.1.3 consists of

- a set $\mathcal{S}_M$ of states over Υ, non-empty subsets $\mathcal{I}_M \subseteq \mathcal{S}_M$ of initial states, and $\mathcal{F}_M \subseteq \mathcal{S}_M$ of final states, satisfying the requirements in Postulate 3.1.2,
- a closed DB-ASM rule r_M over Υ, and
- a binary relation δ_M over $\mathcal{S}_M$ determined by r_M such that

$$\{S_{i+1} \mid (S_i, S_{i+1}) \in \delta_M\} = \{S_i + \Delta \mid \Delta \in \Delta(r_M, S_i)\}$$

  holds.

For the convenience of discussion, we define the substitution for terms.

Definition 3.2.3. Let t be a term, x be a variable, a be a constant and f be a function symbol. Then,

- $x[t/x'] = \begin{cases} t & \text{if } x = x' \\ x & \text{else} \end{cases}$
- $a[t/x'] = a$
- $f(t_1, ..., t_n)[t/x'] = f(t_1[t/x'], ..., t_n[t/x'])$

The substitution of a term t for a variable x in a formula φ (denoted as $\varphi[t/x]$) is defined by the rule of substitution. That is,

- $\varphi[t/x]$ is the result of replacing all free instances of x by t in φ provided that no free variable of t becomes bound after substitution.

Theorem 3.2.1. *Each DB-ASM M defines a database transformation T with the same signature and background as M.*

Proof. We have to show that the five postulates for database transformations are satisfied. As for the sequential time and background postulates (i.e., Postulates 3.1.1 and 3.1.3), these are already built into the definition of a DB-ASM. The same holds for the abstract state postulate (i.e., Postulate 3.1.2) as far as the definition of states is concerned, and the preservation of isomorphisms follows from Lemma 3.2.1. So, we have to concentrate on the bounded exploration and the bounded non-determinism postulates (i.e., Postulates 3.1.4 and 3.1.5).

Regarding bounded exploration we noted above that the assignment rules within the DB-ASM rule r that defines π_M are decisive for the update set $\Delta(r, S)$ for any state S. Hence, if $f(t_1, \dots, t_n) := t_0$ is an assignment occurring within r, and $val_{S,\zeta}(t_i) = val_{S',\zeta}(t_i)$ holds for all $i = 0, \dots, n$ and all variable assignments ζ that have to be considered, then we obtain $\Delta(r, S) = \Delta(r, S')$.

We use this to define a bounded exploration witness $\mathcal{T}_{wit}$. If t_i is ground, we add the access term t_i to $\mathcal{T}_{wit}$. If t_i is not ground, then the corresponding assignment rule must appear within the scope of forall and choice rules introducing the database variables in t_i, as r is closed. Thus, variables in t_i are bound by a a first-order formula φ, i.e. for $fr(t_i) = \{x_1, \dots, x_k\}$ the relevant variable assignments are $\zeta = \{x_1 \mapsto b_1, \dots, x_k \mapsto b_k\}$ with $val_{S,\zeta}(\varphi) = true$. Bringing φ into a form that only uses conjunction, negation and existential quantification with atoms $t_{\beta_i} = t_{\alpha_i}$ $(i = 1, \dots, \ell)$, we can extract a set of access terms $\{(t_{\beta_1}, t_1), \dots, (t_{\beta_\ell}, t_\ell)\}$ such that if S and S' coincide on these access terms, they will also coincide on the formula φ. This is possible, as we evaluate access terms by sets, so conjunction corresponds to union, existential quantification to projection, and negation to building the (finite) complement. We add all the access terms $(t_{\beta_1}, t_{\alpha_1}), \dots, (t_{\beta_\ell}, t_{\alpha_\ell})$ to $\mathcal{T}_{wit}$.

More precisely, if φ is a conjunction $\varphi_1 \wedge \varphi_2$, then $\Delta(r, S_1) = \Delta(r, S_2)$ will hold, if $\{(b_1, \dots, b_k) \mid val_{S_1,\zeta}(\varphi) = true\} = \{(b_1, \dots, b_k) \mid val_{S_2,\zeta}(\varphi) = true\}$ holds (with $\zeta = \{x_1 \mapsto b_1, \dots, x_k \mapsto b_k\}$). If $\mathcal{T}_i$ is a set of access terms such that whenever S_1 and S_2 coincide on $\mathcal{T}_i$, then $\{(b_1, \dots, b_k) \mid val_{S_1,\zeta}(\varphi_i) = true\} = \{(b_1, \dots, b_k) \mid val_{S_2,\zeta}(\varphi_i) = true\}$ will hold $(i = 1, 2)$, then $\mathcal{T}_1 \cup \mathcal{T}_2$ is a set of access terms such that whenever S_1 and S_2 coincide on $\mathcal{T}_1 \cup \mathcal{T}_2$, then $\{(b_1, \dots, b_k) \mid val_{S_1,\zeta}(\varphi) = true\} = \{(b_1, \dots, b_k) \mid val_{S_2,\zeta}(\varphi) = true\}$ will hold.

Similarly, a set of access terms for ψ with the desired property will also be a witness for $\varphi = \neg\psi$, and $\bigcup_{b_{k+1} \in B_{db}} \mathcal{T}_{b_{k+1}}$ with sets of access terms $\mathcal{T}_{b_{k+1}}$ for $\psi[t_{k+1}/x_{k+1}]$ with $val_S(t_{k+1}) = b_{k+1}$ defines a finite set of access terms for $\varphi = \exists x_{k+1}\psi$. In this way, we can restrict ourselves to atomic formulae, which are equations and thus give rise to canonical access terms.

Then by construction, if S and S' coincide on $\mathcal{T}_{wit}$, we obtain $\Delta(r, S) = \Delta(r, S')$. As there are only finitely many assignments rules within r and only finitely many choice and forall rules defining the variables in such assignments, the set $\mathcal{T}_{wit}$ of access terms must be finite, i.e. r satisfies the bounded exploration postulate.

Regarding bounded non-determinism, assuming that M does not satisfy the bounded non-determinism postulate. It means that there does not exist a non-ground access term (t_β, t_α) in $\mathcal{T}_{wit}$. According to our remark above r must contain a choice rule **choose**

$x_1, \dots, x_k$ **with** φ **do** r' **enddo**. Hence, it implies that there exist at least one non-ground access term in $\mathcal{T}_{wit}$ contradicting our assumption. □

3.3 A Characterisation Theorem

In this section we want to show that DB-ASMs capture all database transformations. This constitutes the converse of Theorem 3.2.1, i.e. that every database transformation can be behaviouraly simulated by a DB-ASM. We start with some preliminaries that are only slight adaptations of corresponding definitions and results for the sequential ASM-thesis, except that the term *critical value* has to be defined differently due to the use of variables in access terms. We then show first that a one-step transition from a state to successor states can be expressed by a DB-ASM rule. Here we heavily rely on the abstract state postulate and the bounded non-determinism postulate, which allows us to deal with the restricted non-determinism appropriately.

In a second step we generalise the proof to the complete database transformation using a construction that is similar to the one used in the sequential ASM-thesis. Again, the fact that our bounded exploration witness contains non-ground terms makes up most of the difficulty.

3.3.1 Critical Terms and Critical Elements

Throughout this section we only deal with consistent update sets, which define the progression of states in a run. We now start providing the key link from updates as implied by the state transitions to DB-ASM rules. Same as the previous subsection this is only a slight extension to the work done for the sequential ASM thesis.

Definition 3.3.1. Let $\mathcal{T}_{wit}$ be a bounded exploration witness for the database transformation T. A term that is constructed out of the subterms of $t_\alpha \in \mathcal{T}_{wit}$ and variables $x_1, \dots, x_k$, for which there are access terms $(t_{\beta_1}, t_{\alpha_1}), \dots, (t_{\beta_\ell}, t_{\alpha_\ell}) \in \mathcal{T}_{wit}$ such that $\bigcup_{i=1}^{\ell} fr(t_{\beta_i}) \cup fr(t_{\alpha_i}) = \{x_1, \dots, x_k\}$ holds is called a *critical term*.

From a bounded exploration witness, a set CT of critical terms can be obtained. This definition differs from the one given in [82] in that we consider also non-ground terms. For access terms in $\mathcal{T}_{wit}$ we cannot simply require closure under subterms, as coincidence of structures on $\mathcal{T}_{wit}$ does not carry over to subterms of "associative" access terms (t_β, t_α). Therefore, we need a different approach to define critical elements.

If t_{ct} is a critical term, let $(t_{\beta_1}, t_{\alpha_1}), \dots, (t_{\beta_\ell}, t_{\alpha_\ell})$ be the access terms used in its definition. For a state S choose $b_1, \dots, b_k \in B_{db}$ with $val_{S,\zeta}(t_{\beta_i}) = val_{S,\zeta}(t_{\alpha_i})$ with $\zeta = \{x_1 \mapsto b_1, \dots, x_k \mapsto b_k\}$ for $i = 1, \dots, \ell$, and let $a = val_{S,\{x_1 \mapsto b_1, \dots, x_k \mapsto b_k\}}(t_{ct})$.

Definition 3.3.2. For each state S of a database transformation T, the elements a of $\bar{C}_S$ constructed in this way are called the *critical elements* of S.

Let $C_S = \{val_S(t) \mid t \in \mathcal{T}_{wit}\} \cup \{true, false, \bot\}$ for a state S of the database transformation T. Let

$$B_S = \{a_i \mid f(a_1, \dots, a_n) \in val_S(t_\beta, t_\alpha) \text{ for some access term } (t_\beta, t_\alpha) \in \mathcal{T}_{wit}\}.$$

Let $\bar{C}_S$ denote the *background closure* of $C_S \cup B_S$ containing all complex values that can be constructed out of $C_S \cup B_S$ using the constructors and function symbols (interpreted in S) in Υ_K. Then $\bar{C}_S$ will be the set of critical elements of the state S. The following lemma and its proof are analogous to the result in [82, Lemma 6.2].

Lemma 3.3.1. *For all updates $(f(a_1, \dots, a_n), a_0) \in \Delta(T, S, S')$ and $(S, S') \in \delta_T$ the values $a_0, \dots, a_n$ are critical elements of S.*

Proof. Assume one of the a_i is not critical. Then choose a structure S_1 by replacing a_i with a fresh value b without changing anything else. Thus, S_1 is a state isomorphic to S by the abstract state postulate.

Let (t_β, t_α) be an access term in $\mathcal{T}_{wit}$. Then we must have $val_S(t_\beta, t_\alpha) = val_{S_1}(t_\beta, t_\alpha)$, so S and S_1 coincide on $\mathcal{T}_{wit}$. From the bounded exploration postulate we obtain $\Delta(T, S) = \Delta(T, S_1)$ and thus $(f(a_1, \dots, a_n), a_0) \in \Delta(T, S_1, S_1')$ for some $(S_1, S_1') \in \delta_T$.

However, a_i does not appear in the structure S_1, and hence cannot appear in S_1' either, nor in $\Delta(T, S_1, S_1')$, which gives a contradiction. □

3.3.2 Rules for One-Step Updates

In [82] it is a straightforward consequence of Lemma 6.2 that individual updates can be represented by assignments rules, and consistent update sets by par-blocks of assignment rules. In our case showing that $\Delta(T, S)$ can be represented by a DB-ASM rule requires a bit more work, which relies heavily on the abstract state postulate and the bounded non-determinism postulate. We address this in the next lemma.

Lemma 3.3.2. *Let T be a database transformation. For every state $S \in \mathcal{S}_T$ there exists a rule r_S such that $\Delta(T, S) = \Delta(r_S, S)$, and r_S only uses critical terms.*

Proof. $\Delta(T,S)$ is a set of update sets. Let $\{S_1,\dots,S_m\} = \{S' \mid (S,S') \in \delta_T\}$. Then $\Delta(T,S) = \{\Delta(T,S,S_i) \mid 1 \le i \le m\}$.

Now consider any update $u = (f(a_1,\dots,a_n),a_0) \in \Delta(T,S,S_i)$ for some $i \in \{1,\dots,m\}$. According to Lemma 3.3.1 the values $a_0,\dots,a_n$ are critical and hence representable by terms involving variables from access terms in $\mathcal{T}_{wit}$, i.e. $a_i = val_{S,\zeta}(t_i)$ with either $fr(t_i) \subseteq \{x_1,\dots,x_k\}, \zeta = \{x_1 \mapsto b_1,\dots,x_k \mapsto b_k\}$ and

$$(b_1,\dots,b_k) \in \mathcal{B}_u = \{(b_1,\dots,b_k) \in B_{db}^k \mid \bigwedge_{1\le i\le \ell} val_{S,\zeta}(t_{\beta_i}) = val_{S,\zeta}(t_{\alpha_i})\}$$

with access terms $(t_{\beta_i},t_{\alpha_i}) \in \mathcal{T}_{wit}$ $(i = 1,\dots,\ell)$ and $fr(t_{\beta_i}) \subseteq \{x_1,\dots,x_k\}$, or t_i is a ground critical term.

Therefore, we distinguish two cases:

I. At least one of the terms $t_0,\dots,t_n$ is not a ground term.

II. All terms $t_0,\dots,t_n$ are ground terms.

Case I. We first assume that none of terms $t_0,\dots,t_n$ contain location operators. The access terms $(t_{\beta_i},t_{\alpha_i})$ define a finite set of locations

$$\begin{aligned}\mathcal{L} \;=\; \{f(a_1,\dots,a_n) \mid a_i = val_{S,\zeta}(t_i) \text{ for } i = 1,\dots,n, \text{ and} \\ \zeta = \{x_1 \mapsto b_1,\dots,x_k \mapsto b_k\} \text{ for } (b_1,\dots,b_k) \in \mathcal{B}_u\}.\end{aligned}$$

However, instead of looking at updates at these locations we switch to a relational perspective, i.e. we replace $f \in \Upsilon$ with arity n by a relation symbol R^f of arity $n+1$, so $f_S(a_1,\dots,a_n) = a_0$ holds iff $R^f_S(a_1,\dots,a_n,a_0) = true$. A non-trivial update $u = (f(a_1,\dots,a_n),a_0)$ is accordingly represented by two relational updates

$$u_d = (R^f(a_1,\dots,a_n,f_S(a_1,\dots,a_n)),false) \text{ and } u_i = (R^f(a_1,\dots,a_n,a_0),true).$$

So, instead of locations in $\mathcal{L}$ we consider locations in $\mathcal{L}_{pre} \cup \mathcal{L}_{post}$ with

$$\begin{aligned}\mathcal{L}_{pre} = \{R^f(a_1,\dots,a_n,a_0) \mid a_i = val_{S,\zeta}(t_i) \text{ for } 1 \le i \le n, \\ a_0 = val_{S,\zeta}(f(t_1,\dots,t_n)) \text{ for } \zeta = \{x_1 \mapsto b_1,\dots,x_k \mapsto b_k\} \\ \text{and } (b_1,\dots,b_k) \in \mathcal{B}_u\}\end{aligned}$$

and

$$\mathcal{L}_{post} = \{R^f(a_1, \dots, a_n, a_0) \mid a_i = val_{S,\zeta}(t_i) \text{ for } 0 \leq i \leq n \text{ for } \zeta = \{x_1 \mapsto b_1, \dots, x_k \mapsto b_k\} \text{ and } (b_1, \dots, b_k) \in \mathcal{B}_u\}.$$

Furthermore, we may assume that the set $\mathcal{B}_u$ is minimal in the sense that we may not find additional access terms that would define a subset $\mathcal{B}'_u \subsetneq \mathcal{B}_u$ still containing the value tuple $(b_1, \dots, b_k)$ that is needed to define the update u.

Then each tuple $(a_1, \dots, a_n, a_0) \in \mathcal{L}_{pre} \cup \mathcal{L}_{post}$ defines a substructure of S with base set $B' = \{a_0, \dots, a_n, true, false\}$ and all functions (in fact: relations) restricted to this base set. In doing so all substructures defined by $\mathcal{L}_{pre}$ (and analogously by $\mathcal{L}_{post}$) are pairwise equivalent, and the induced isomorphisms are defined by permutations of tuples in $\mathcal{B}_u$. If they were not equivalent, we could find a distinguishing access term $(t_{\beta_{\ell+1}}, t_{\alpha_{\ell+1}}) \in \mathcal{T}_{wit}$ that would define a subset $\mathcal{B}'_u \subsetneq \mathcal{B}_u$ thereby violating the minimality assumption for $\mathcal{B}_u$. Let E_{pre} and E_{post} denote these equivalence classes of substructures, respectively.

If $\ell' \in \mathcal{L}$ is not updated in S_i – hence, corresponding locations in $\mathcal{L}_{pre}$ and $\mathcal{L}_{post}$ are neither updated – then the substructures $S_{\ell'}$ in E_{pre} (and E_{post}, respectively) that are defined by ℓ' are preserved in S_i, i.e. $S_{\ell'} \preceq S_i$. From Lemma 3.1.2 we can get that $\varsigma(\Delta(T,S)) = \Delta(T,\varsigma(S)) = \Delta(T,S)$ for the case that ς is an automorphism of S. It means that for an update in $\Delta(T,S,S_j)$ there is a translated update (by means of ς) in $\Delta(T,S,\varsigma(S_j))$. Then we conclude that for all S_j $(j = 1, \dots, m)$ there is some $\ell'' \in \mathcal{L}$ with $S_{\ell''} \preceq S_j$, and each $S_{\ell''} \in E_{pre}$ (and $S_{\ell''} \in E_{post}$, respectively) is preserved in some S_j. In particular, there is also some successor state S_j of S with $S_\ell \preceq S_j$ for $u = (\ell, a_0)$.

Thus, we obtain two subcases:

1) If ℓ is updated in all $S_1, \dots, S_m$, i.e. there exist values $a_0^1, \dots, a_0^m$ with $(\ell, a_0^i) \in \Delta(T,S,S_i)$ for all $i = 1, \dots, m$, then all $\ell' \in \mathcal{L}$ are also updated in all S_i. If (ℓ, a_0^i) is represented by the assignment rule $f(t_1, \dots, t_n) := t_0^i$ with $x_1, \dots, x_k$ interpreted by $(b_1, \dots, b_k) \in \mathcal{B}_u$, then the fact that almost all substructures defined by these interpretations of $t_1, \dots, t_n$ and any value other than a_0^i are preserved – and hence by virtue of Lemma 3.1.2 as we explained before equivalent substructures are preserved in the other S_j – implies that each instantiation of the rule $f(t_1, \dots, t_n) := t_0^i$ with values from $\mathcal{B}_u$ defines an update in one of the update sets $\Delta(T,S,S_j)$. Hence these updates can be collectively represented by the rule

$$\textbf{choose } x_1, \dots, x_k \textbf{ with } t_{\beta_1}(\overline{x}_1) = t_{\alpha_1}(\overline{x}_1) \wedge \dots \wedge t_{\beta_\ell}(\overline{x}_\ell) = t_{\alpha_\ell}(\overline{x}_\ell)$$

do $f(t_1, \dots, t_n) := t_0^i$ **enddo**

Here the $\overline{x}_1, \dots, \overline{x}_\ell$ denote tuples of variables among $x_1, \dots, x_k$ appearing in $t_{\beta_1}, \dots, t_{\beta_\ell}$, respectively. In case all the terms t_0^i for $i = 1, \dots, m$ are identical (to say t_0), we obtain in fact the rule $r_S^{(u)}$ as

choose $x_1, \dots, x_k$ **with** $t_{\beta_1}(\overline{x}_1) = t_{\alpha_1}(\overline{x}_1) \wedge \dots \wedge t_{\beta_\ell}(\overline{x}_\ell) = t_{\alpha_\ell}(\overline{x}_\ell)$
do $f(t_1, \dots, t_n) := t_0$ **enddo**

In general, however, it is possible that different terms t_0^i must be chosen, so all updates to locations in $\mathcal{L}$ are represented by the rule $r_S^{(u)}$, which becomes

choose $x_1^{(1)}, \dots, x_k^{(1)}, \dots, x_1^{(m)}, \dots, x_k^{(m)}$
with $\bigwedge_{1 \le j_1 < j_2 \le m} (x_1^{(j_1)}, \dots, x_k^{(j_1)}) \neq (x_1^{(j_2)}, \dots, x_k^{(j_2)})$
$\wedge \bigwedge_{1 \le j \le m} t_{\beta_1}(\overline{x}_1^{(j)}) = t_{\alpha_1}(\overline{x}_1^{(j)}) \wedge \dots \wedge t_{\beta_\ell}(\overline{x}_\ell^{(j)}) = t_{\alpha_\ell}(\overline{x}_\ell^{(j)})$
do
par
$f(t_1, \dots, t_n)[x_1^{(1)}/x_1, \dots, x_k^{(1)}/x_k] := t_0^1[x_1^{(1)}/x_1, \dots, x_k^{(1)}/x_k]$
$\vdots \qquad \vdots$
$f(t_1, \dots, t_n)[x_1^{(m)}/x_1, \dots, x_k^{(m)}/x_k] := t_0^m[x_1^{(m)}/x_1, \dots, x_k^{(m)}/x_k]$
endpar
enddo

2) If only ℓ is updated in S_i, but no other $\ell' \in \mathcal{L}$ is, then only one $\ell' \in \mathcal{L}$ is updated in each S_j for $j = 1, \dots, m$. Analogously, if $m - i$ locations in $\mathcal{L}$ are not updated, then i locations in $\mathcal{L}$ are updated in each S_j for $j = 1, \dots, m$. Using exactly the same arguments as in case 1), we now can represent these updates collectively by the rule $r_S^{(u)}$, which now becomes

choose $x_1^{(1)}, \dots, x_k^{(1)}, \dots, x_1^{(i)}, \dots, x_k^{(i)}$
with $\bigwedge_{1 \le j_1 < j_2 \le i} (x_1^{(j_1)}, \dots, x_k^{(j_1)}) \neq (x_1^{(j_2)}, \dots, x_k^{(j_2)})$
$\wedge \bigwedge_{1 \le j \le i} t_{\beta_1}(\overline{x}_1^{(j)}) = t_{\alpha_1}(\overline{x}_1^{(j)}) \wedge \dots \wedge t_{\beta_\ell}(\overline{x}_\ell^{(j)}) = t_{\alpha_\ell}(\overline{x}_\ell^{(j)})$
do

par

$f(t_1,\dots,t_n)[x_1^{(1)}/x_1,\dots,x_k^{(1)}/x_k] := t_0^1[x_1^{(1)}/x_1,\dots,x_k^{(1)}/x_k]$

$\vdots \qquad \vdots$

$f(t_1,\dots,t_n)[x_1^{(i)}/x_1,\dots,x_k^{(i)}/x_k] := t_0^i[x_1^{(i)}/x_1,\dots,x_k^{(i)}/x_k]$

endpar

enddo

By exploiting Lemma 3.1.2, we showed how to create a proper choice rule with respect to $\mathcal{B}_u$ that is minimal. However, the created choice rule does not capture all update sets in $\Delta(T,S)$. If there exists another update in an update set that is not in the orbit of $\Delta(T,S,S_i)$ $(1 \leq i \leq m)$ under ς, we can use the same argument to obtain another choice rule. As the orbits are disjoint, we end up with a choice of choice rules, which can be combined into a single choice rule. The underlying condition for constructing such a single choice rule is the finiteness of $\Delta(T,S)$, i.e., there are only finitely many update sets created by T over state S. This can be assured by Lemma 3.3.1 and the bounded non-determinism postulate. It is sufficient to consider the subset $CT' \subseteq CT$ defined by closed terms in $\mathcal{T}_{wit}$ and their subterms and tuple terms defined by some set of access terms in $\mathcal{T}_{wit}$. Hence, as $\mathcal{T}_{wit}$ is finite, CT' is also finite, and consequently there can only be finitely many branches from the database part of the state S, which is a finite structure.

Now we revise the previous assumption that none of terms $t_0,\dots,t_n$ contain location operators to a general case, i.e., location operators may appear in the terms $t_0,...,t_n$ of an assignment rule $f(t_1,...,t_n) := t_0$. Let f_ℓ be a unary function symbol such that $x_{t_i} = f_\ell(i)$, then, without loss of generality, we can replace the terms $t_1,...,t_n$ of an assignment rule $f(t_1,...,t_n) := t_0$ with the variables $x_{t_1},...,x_{t_n}$, such that

par

$x_{t_1} := t_1$

$\vdots$

$x_{t_n} := t_n$

endpar;

$f(x_{t_1},...,x_{t_n}) := t_0$

It means that we can simplify the construction of rules for updates which may correspond to terms with location operators by only considering the case that location operators appear at the right hand side of an assignment rule. If a term t_i ($i \in [1,n]$) at the left hand side contains a location operator, by the above translation, we may treat it as being a term at the right hand side of another assignment rule again.

Suppose that the outermost function symbol of term t_0 is a location operator ρ, e.g.,$t_0 = \rho(m)$ where $m = \{\!\{t_0^{'}|$ for all values $\overline{a} = (a_1, ..., a_p)$ in $\overline{y} = (y_1, ..., y_p)$ such that $val_{S,\zeta[x_1 \mapsto b_1,...,x_k \mapsto b_k]}(\varphi(\overline{x}, \overline{a})) = true\}\!\}$, and $\overline{x}$ denotes a tuple of variables among $x_1, ..., x_k$. Then for each assignment rule $f(x_{t_1}, ..., x_{t_n}) := t_0$ in which t_0 contains a location operator as described before, we can construct the following rule to remove the location operator ρ by a let rule and a forall rule:

> **let** $\theta(f(x_{t_1}, ..., x_{t_n})) = \rho$ **in**
>
> **forall** $y_1,...,y_p$ **with** $\varphi(\overline{x}, \overline{y})$
>
> **do**
>
> $f(x_{t_1}, ..., x_{t_n}) := t_0^{'}$
>
> **enddo**;
>
> **endlet**

This construction can be conducted iteratively. If the outermost function symbol of the above term $t_0^{'}$ is a location operator, then we need to construct a rule in a similar way to replace the assignment rule $f(x_{t_1}, ..., x_{t_n}) := t_0^{'}$. This procedure continues until the right hand side of an assignment rule is a term without any location operator.

Case II. In case of a simple update $f(t_1, \dots, t_n) := t_0$ without free variables we consider the substructure defined by $\{a_1, \dots, a_n, val_S(f(t_1, \dots, t_n)), true, false\}$ as before. However, in this case it is the only substructure in its equivalence class. Furthermore, the substructure can be represented by ground access terms.

According to the bounded non-determinism postulate, we know that, if the location $\ell = f(a_1, \dots, a_n)$ were not updated in some S_j with $j \in \{1, \dots, m\}$, the same would apply for all S_i ($i = 1, \dots, m$) contradicting the fact that the update (ℓ, a_0) appears in one S_i. Consequently, $(f(a_1, \dots, a_n), a_0) \in \Delta(T, S, S_i)$ for all $i = 1, \dots, m$, and these updates can be collectively represented by the simple assignment rule $r_S^{(u)}$, which now becomes

$$f(t_1, \dots, t_n) := t_0.$$

Finally, we construct r_S by using the **par**-construct:

$$r_S = \textbf{par } r_S^{(u_1)} \dots r_S^{(u_p)} \textbf{ endpar}$$

for $\{u_1, \dots, u_p\} = \bigcup_{i=1}^{m} \Delta(T, S, S_i)$.

□

3.3.3 Rules for Multiple-Step Updates

Let us now extend Lemma 3.3.2 to the construction of a DB-ASM rule that captures the complete behaviour of a database transformation T. For this fix a bounded exploration witness $\mathcal{T}_{wit}$ and the set CT of critical terms derived from it. Furthermore, for a state S of T fix the rule r_S as in Lemma 3.3.2.

For $t_{ct} \in CT$, let $(t_{\beta_1}, t_{\alpha_1}), \dots, (t_{\beta_\ell}, t_{\alpha_\ell})$ be the access terms in $\mathcal{T}_{wit}$ defining $fr(t_{ct}) = \{x_1, \dots, x_k\}$. For a state $S \in \mathcal{S}_T$ define

$$val_S(t_{ct}) = \{val_{S,\zeta}(t_{ct}) \mid \zeta = (x_1 \mapsto b_1, \dots, x_k \mapsto b_k) \text{ and } \bigwedge_{1 \le i \le \ell} val_{S,\zeta}(t_{\beta_i}) = val_{S,\zeta}(t_{\alpha_i})\}.$$

The following two lemmata extend Lemma 3.3.2 first to states that coincide with S on critical terms, then to isomorphic states.

Lemma 3.3.3. *Let $S, S' \in \mathcal{S}_T$ be states that coincide on the set CT of critical terms. Then $\Delta(r_S, S') = \Delta(T, S')$ holds.*

Proof. As S and S' coincide on CT, they also coincide on $\mathcal{T}_{wit}$, which gives $\Delta(T, S) = \Delta(T, S')$ by the bounded exploration postulate. Furthermore, we have $\Delta(r_S, S) = \Delta(T, S)$ by Lemma 3.3.2. As r_S uses only critical terms, the updates produced in state S must be the same as those produced in state S', i.e. $\Delta(r_S, S) = \Delta(r_S, S')$, which proves the lemma. □

Lemma 3.3.4. *Let S, S_1, S_2 be states with S_1 isomorphic to S_2 and $\Delta(r_S, S_2) = \Delta(T, S_2)$. Then also $\Delta(r_S, S_1) = \Delta(T, S_1)$ holds.*

Proof. Let ς denote an isomorphism from S_1 to S_2. Then $\Delta(r_S, S_2) = \varsigma(\Delta(r_S, S_1))$ holds by Lemma 3.1.2, and the same applies to $\Delta(T, S_2) = \varsigma(\Delta(T, S_1))$. As we presume $\Delta(r_S, S_2) = \Delta(T, S_2)$, we obtain $\varsigma(\Delta(r_S, S_1)) = \varsigma(\Delta(T, S_1))$ and hence $\Delta(r_S, S_1) = \Delta(T, S_1)$, as ς is an isomorphism. □

Next, in the spirit of [82] we want to extend the equality of sets of update sets for T and r_S to a larger class of states by exploiting the finiteness of the bounded exploration witness $\mathcal{T}_{wit}$. For this, we define the notion of T-equivalence similar to the corresponding notion for the sequential ASM thesis, with the difference being that in our case we cannot take $\mathcal{T}_{wit}$, but must base our definition and the following lemma on CT.

Definition 3.3.3. States $S, S' \in \mathcal{S}_T$ are called *T-similar* iff $E_S = E_{S'}$ holds, where E_S is an equivalence relation on CT defined by

$$E_S(t_{ct_1}, t_{ct_2}) \Leftrightarrow val_S(t_{ct_1}) = val_S(t_{ct_2}).$$

Lemma 3.3.5. *We have $\Delta(r_S, S') = \Delta(T, S')$ for every state S' that is T-similar to S.*

Proof. Replace every element in S' that also belongs to S by a fresh element. This defines a structure S_1 isomorphic to S' and disjoint from S. By the abstract state postulate S_1 is a state of T. Furthermore, by construction S_1 is also T-similar to S' and hence also to S.

Now define a structure S_2 isomorphic to S_1 such that $val_{S_2}(t_{ct}) = val_S(t_{ct})$ holds for all critical terms $t_{ct} \in CT$. This is possible, as S and S_1 are T-similar, i.e. we have $val_S(t_{ct_1}) = val_S(t_{ct_2})$ iff $val_{S_1}(t_{ct_1}) = val_{S_1}(t_{ct_2})$ for all critical terms t_{ct_1}, t_{ct_2}. By the abstract state postulate S_2 is also a state of T.

Using Lemma 3.3.3 we conclude $\Delta(r_S, S_2) = \Delta(T, S_2)$, and by Lemma 3.3.4 we obtain $\Delta(r_S, S') = \Delta(T, S')$ as claimed. □

We are now able to prove our main result, first generalising Lemma 3.3.2 to multiple-step updates in the next lemma, from which the proof of the main characterisation theorem is straightforward.

Lemma 3.3.6. *Let T be a database transformation with signature Υ. Then there exists a DB-ASM rule r over Υ, with same background as T such that $\Delta(r, S) = \Delta(T, S)$ holds for all states $S \in \mathcal{S}_T$.*

Proof. In order to decide whether equivalence relations E_S and $E_{S'}$ coincide for states $S, S' \in \mathcal{S}_T$ it is sufficient to consider the subset $CT' \subseteq CT$ defined by closed terms in $\mathcal{T}_{wit}$ and their subterms and tuple terms $(x_1, \dots, x_k) \in CT$ that are defined by some set of access terms in $\mathcal{T}_{wit}$. Hence, as $\mathcal{T}_{wit}$ is finite, CT' is also finite, and by the bounded non-determinism postulate there can only be finitely many such equivalence relations. Let these be $E_{S_1}, \dots, E_{S_n}$ for states $S_1, \dots, S_n \in \mathcal{S}_T$.

For $i = 1, \dots, n$ we construct first-order formulae φ_i such that $val_S(\varphi_i) = true$ holds iff S is T-similar to S_i. For this let $CT' = \{t_{ct_1}, \dots, t_{ct_m}\}$, and define terms

$$\bar{t}_{ct_j} = \begin{cases} t_{ct_j} & \text{if } t_{ct_j} \text{ is closed} \\ \langle (x_1, \dots, x_k) \mid \bigwedge\limits_{1 \le i \le \ell} t_{\beta_i} = t_{\alpha_i} \rangle & \text{if } t_{ct_j} = (x_1, \dots, x_k) \text{ with variables taken} \\ & \text{from } (t_{\beta_1}, t_{\alpha_1}), \dots, (t_{\beta_\ell}, t_{\alpha_\ell}) \end{cases}$$

exploiting the fact that the background structures provide constructors for multisets and pairs (and thus also tuples). Then

$$\varphi_i = \bigwedge_{\substack{1 \le j_1, j_2 \le m \\ E_{S_i}(t_{ct_{j_1}}, t_{ct_{j_2}})}} \bar{t}_{ct_{j_1}} = \bar{t}_{ct_{j_2}} \quad \wedge \bigwedge_{\substack{1 \le j_1, j_2 \le m \\ \neg E_{S_i}(t_{ct_{j_1}}, t_{ct_{j_2}})}} \bar{t}_{ct_{j_1}} \neq \bar{t}_{ct_{j_2}}$$

asserts that $E_S = E_{S_i}$ holds. Now define the rule r by

par **if** φ_1 **then** r_{S_1} **endif**
if φ_2 **then** r_{S_2} **endif**
⋮
if φ_n **then** r_{S_n} **endif**
endpar

If $S \in \mathcal{S}_T$ is any state of T, then S is T-equivalent to exactly one S_i $(1 \le i \le n)$, which implies $val_S(\varphi_j) = true$ iff $j = i$, and hence $\Delta(r, S) = \Delta(r_{S_i}, S) = \Delta(T, S)$ by Lemma 3.3.5. □

Theorem 3.3.1. *For every database transformation T there exists an equivalent DB-ASM M.*

Proof. By Lemma 3.3.6 there is a DB-ASM rule r with $\Delta(r, S) = \Delta(T, S)$ for all $S \in \mathcal{S}_T$. Define M with the same signature and background as T (and hence $\mathcal{S}_M = \mathcal{S}_T$), $\mathcal{I}_M = \mathcal{I}_T$, $\mathcal{F}_M = \mathcal{F}_T$, and program $\delta_M = r$. □

Note that for the proof of Theorem 3.3.1 we constructed a DB-ASM rule that does not use sequence nor let rules, so by Theorem 3.2.1 these two constructions can be considered to be merely "syntactic sugar". However, as discussed before the let rules are actually more than that, as they capture aggregate updates that exploit parallelism. Whether this can be extended to capture various aspects of parallelism, thus looking deeper inside database transformations, is an open problem.

Chapter 4

Relational and XML Database Transformations

In this chapter, we discuss database transformations over relational and XML databases.

Relational databases are widely recognised as an efficient technology for data management. The first task of this chapter is to specify a computational model for database transformations on relational databases. The approach we adopt here is to customise the background postulate with a relational background. In particular, we exploit relational algebra for organising query execution and optimisation at an internal implementation level, in terms of the constructivity of a relational background. It turns out that DB-ASMs with relational backgrounds characterise relational database transformations. Every relational database transformation (i.e. a database transformation over a relational database satisfying the five postulates) can be behaviourally simulated by a DB-ASM with a relational background and vice versa. We also identify and compare several subclasses of relational database transformations characterised by such relational backgrounds.

For XML databases most current approaches that deal with XML documents have a fixed abstraction level. It means that each node or edge of interest must be explicitly identified in a database transformation. In order to accommodate the diversity of user requirements by providing a more intuitive, user-friendly and less implementation-specific computation model, manipulations on XML trees at an arbitrary flexible abstraction level would be desirable. It is appealing to have manipulations such as deleting, modifying or inserting subtrees, copying and replacing contexts, etc. on portions of a tree. Therefore, the second task of this chapter is to investigate XML database transformations with an emphasis on structured data updates. This task naturally brings up the following two questions:

- What may be an appropriate computation model for XML database transformations at a flexible abstraction level?
- Is the class of algorithms captured by this computation model the same as XML database transformations stipulated by the postulates with the same signature and background?

To answer these questions, we develop a tree-based background that customises the background postulate for XML database transformations at a flexible abstraction level. Then we exploit weak Monadic Second-Order (MSO) logic as a means to define an alternative, more elegant way of expressing XML database transformations. It is revealed that incorporating MSO logic into DB-ASM rules can not actually increase the expressiveness of XML machines.

The rest of the chapter is organised as follows. We discuss relational backgrounds and relational database transformations in Section 4.1. In Section 4.2 we formally define XML trees, XML contexts and a tree algebra. They give rise to type constructors and functions that must be considered as part of a tree-based background class. In Section 4.3, weak MSO logic is introduced, which adds further requirements to tree-based backgrounds. The final definition of tree-based backgrounds, which also comprise tree type schemes in order to capture schema information based on EDTDs, is presented in Section 4.4. Linking tree-based backgrounds with DB-ASMs gives the first computational model for XML database transformations. The alternative computational model of XML machines which exploit weak MSO logic is introduced in Section 4.5. The proof for the main result that the model of XML machines is equivalent to the DB-ASM model with tree-based backgrounds is presented in Subsection 4.5.2.

4.1 Relational Database Transformations

Relational database transformations are the most ubiquitous class of database transformations, in which states are formalised on the basis of relations. In this section, we address two issues concerning relational database transformations: (i) to clarify computation backgrounds specific to relational database transformations, which we call relational backgrounds; (ii) to investigate how relational backgrounds interact with relational schemata, and thereby impose restrictions on the underlying states.

4.1.1 Relational Model

The Relational Data Model (RDM) was proposed by Edgar Codd [56, 57]. Let $\mathcal{D} = \{D_i\}_{i \in I}$ be a family of atomic domains containing all kinds of possible values. Then a *relation schema* R consists of a finite, non-empty set $attr(R)$ of *attributes* and a *domain assignment* $dom : attr(R) \rightarrow \mathcal{D}$. A *tuple* over a relation schema R is a mapping

$$\alpha_t : attr(R) \rightarrow \bigcup_{D_i \in \mathcal{D}} D_i$$

with $\alpha_t(A) \in dom(A)$ for all $A \in attr(R)$. A *relation* $I(R)$ over a relation schema R is a finite set of tuples over R. A *relational database schema* $\mathfrak{R}$ is a finite, non-empty set of relation schemata, and a *relational database instance* $I(\mathfrak{R})$ over a relational database schema $\mathfrak{R}$ assigns to each $R \in \mathfrak{R}$ a relation $I(R)$ over R.

First normal form is an important criterion for a relation schema to be well-designed. It restricts that attributes in any relation schema may only be associated with atomic values. A relaxation of this condition naturally generalises the RDM to the Nested Relational Data Model (NRDM), which permits relations to be the values of *relation-valued attributes*. With the presence of relation-valued attributes, nest and unnest operations are indispensable in the NRDM. They serve to shift from a relation to its subrelations, or vice versa. The NRDM was first formulated by Makinouchi [110], and it has since been extensively investigated in various works, e.g. [141, 123, 105].

A *nested relation schema* R is inductively defined to be a pair $(attr(R), dom)$ of a finite, non-empty set $attr(R)$ of attributes and a domain assignment dom such that

- $attr(R)$ consists of
  - a set $\{A_1, ..., A_n\}$ of *atomic attributes*, and
  - a set $\{R_1, ..., R_m\}$ of *relation-valued attributes* that are nested relation schemata,
- dom associates an atomic domain $dom(A_i) \in \mathcal{D}$ to each $A_i \in attr(R)$, and a schema domain $dom(R_j)$ to each $R_j \in attr(R)$, where the definition of schema domain is as follows.

Let $\mathcal{P}(D)$ denote the power set of D. A *schema domain* $dom(R)$ over a nested relation schema R is recursively defined by

- For $R = \{A_1, ..., A_k\}$, $dom(R) = \mathcal{P}(dom(A_1) \times ... \times dom(A_k))$;

- For $R = \{A_1, ..., A_n, R_1, ..., R_m\}$, $dom(R) = \mathcal{P}(dom(A_1) \times ... \times dom(A_n) \times dom(R_1) \times ... \times dom(R_m))$.

Let R be a nested relation schema. A *tuple* over R is a mapping

$$\alpha_t : attr(R) \rightarrow \bigcup_{A \in attr(R)} dom(A) \cup \bigcup_{R' \in attr(R)} dom(R')$$

with $\alpha_t(A) \in dom(A)$ for each atomic attribute $A \in attr(R)$ and $\alpha_t(R') \in dom(R')$ for each relation-valued attribute $R' \in attr(R)$. A *nested relation* $I(R)$ over R is a finite set of tuples over R. A *nested relational database schema* $\mathfrak{R}$ is a finite, non-empty set of nested relation schemata, and a *nested relational database instance* $I(\mathfrak{R})$ over a nested relational database schema $\mathfrak{R}$ assigns to each $R \in \mathfrak{R}$ a nested relation $I(R)$ over R.

A fundamental assumption of relational models is that data is structured as relations (i.e., as subsets of the cartesian product of certain domains). Therefore, both the RDM and the NRDM are varieties of relational models. The RDM is clearly just a special case of the NRDM in which there are no relation-valued attributes.

4.1.2 Relational Background Class

A relational background class provides all possible elements which can take part in a relational database transformation together with all operations that can be used to enact transformations on these elements. In addition to this, a relational type scheme controls the way in which states of a relational database transformation are instantiated.

We start by defining the signature of a relational background.

Definition 4.1.1. A *relational background signature* Υ_K^{rel} comprises

- type constructors for finite tuple $(\cdot)$, finite set $\{\cdot\}$ and binary label $:$,
- a finite set of static function symbols containing at least $\emptyset$ (empty set), $\{x\}$ (singleton set), $\cup$ (set union), and $\in$ (set membership) to deal with set-based operations, and
- type constructors and function symbols as requested in Postulate 3.1.3.

The set $\mathcal{D}$ of atomic domains in a relational model contains a finite, non-empty set of attribute names, and by default $\mathcal{D}$ also contains the Boolean domain, i.e., *Bool*. The type constructors $\{\cdot\}$, $(\cdot)$ and $:$ are used in a rather restricted but mixed way to construct relational background structures.

Definition 4.1.2. A *relational background class* K^{rel} consists of

- a relational background signature Υ_K^{rel},
- a set $\mathcal{D}$ of atomic domains, and
- a *relational background structure* with Υ_K^{rel} over $\mathcal{D}$ as defined in Definition 3.1.7.

4.1.3 Relational Type Schemes

The basic building blocks of a relational type scheme are *atomic value types* (e.g., Integer, String, Date, Bool) which are called atomic domains in a relational model. Without loss of generality, we will use $\mathcal{D}$ to refer to both a family of atomic value types in a relational type scheme and a set of atomic domains in a relational model, depending on the context.

A relational type scheme has two type constructors: tuple $(\cdot)$ and set $\{\cdot\}$. Let $\mathcal{A}$ be a finite set of attributes consisting of a subset $\mathcal{A}_A \subseteq \mathcal{A}$ of atomic attributes and a subset $\mathcal{A}_R \subseteq \mathcal{A}$ of relation-valued attributes. Furthermore, $\mathcal{A}_A \cap \mathcal{A}_R = \emptyset$. Then, a type in a relational type scheme over $\mathcal{D}$ can be inductively defined by

- $\tau :\equiv A_A : D | A_R : \tau_R$, and
- $\tau_R :\equiv \{(\tau_1, ..., \tau_k)\}$,

where $A_A \in \mathcal{A}_A$, $A_R \in \mathcal{A}_R$ and $D \in \mathcal{D}$.

The *interpretation* of a type τ (expressed by $[\![\tau]\!]$) is a collection of values, which are either atomic or complex, inductively defined by

- $[\![A_A : D]\!] = A_A : D$,
- $[\![A_R : \tau_R]\!] = A_R : [\![\tau_R]\!]$,
- $[\![\{(\tau_1, ..., \tau_k)\}]\!] = \mathcal{P}(\bigcup_{a_i \in [\![\tau_i]\!], i=1,...,k} \{(a_1, \ldots, a_k)\})$.

There are two kinds of label types: $A_A : D$ labelled as a type for an atomic attribute and $A_R : \tau_R$ labelled as a type for a relation-valued attribute. Correspondingly, the interpretation of type $A_A : D$ is a set of values from atomic domains with a label A_A, while the interpretation of type $A_R : \tau_R$ is a set of values from the interpretation of type τ_R with a label A_R. A type $\{(\tau_1, ..., \tau_k)\}$ is interpreted as a set of relations that have attributes of types $\tau_1, ..., \tau_k$ in order.

Definition 4.1.3. Let $\mathfrak{R}$ be a set of relation symbols in the signature of a state. Then a *relational type scheme* is a pair $(\mathfrak{T}, typ)$ consisting of

- a finite, non-empty set $\mathfrak{T}$ of types defined by the above system, and
- a type assignment $typ : \mathfrak{R} \to \mathfrak{T}$ that associates a type $typ(R) \in \mathfrak{T}$ to each relation symbol $R \in \mathfrak{R}$.

Database transformations are normally considered to be associated with a pair of database schemata: input and output database schemata. These schemata are reflected into a relational background via a relational type scheme as follows. All dynamic relational symbols in the state signature are categorised into three kinds: *in*, *out* and *temp*, corresponding to relational symbols in an input database schema, in an output database schema and additionally auxiliary relational symbols, respectively. That is, $\mathfrak{R} = \mathfrak{R}_{in} \cup \mathfrak{R}_{out} \cup \mathfrak{R}_{temp}$. By defining a relational type scheme, each relational symbol is restricted to a specific type that remains fixed during a database transformation and a (nested) relation schema is represented by such a type. $\mathfrak{R}_{out}$ and $\mathfrak{R}_{temp}$ may be empty and $\mathfrak{R}_{in}$ contains at least one relational symbol. Relational symbols in $\mathfrak{R}_{temp}$ are used for facilitating computations and exist only temporarily in intermediate stages of a database transformation.

Example 4.1.1. Consider an input database schema $\mathfrak{R}_{in} = \{R_1\}$ where $R_1 = \{A_1 : \{A_{11} : D_{11}, A_{12} : D_{22}\}, A_2 : D_2, A_3 : D_3\}$ and an output database schema $\mathfrak{R}_{out} = \{R_1, R_2\}$ where $R_2 = \{A_2 : D_2\}$. A database transformation starting from $I(\mathfrak{R}_{in})$ and ending at $I(\mathfrak{R}_{out})$ may have two relational type schemes $(\mathfrak{T}_1, typ_1)$ and $(\mathfrak{T}_2, typ_2)$ associated with input and output database schemata, respectively, such that

- $typ_1(R_1) = \tau_1$, $typ_2(R_1) = \tau_1$ and $typ_2(R_2) = \tau_2$, and
  - $\tau_1 = \{(A_1 : \{(A_{11} : D_{11}, A_{12} : D_{22})\}, A_2 : D_2, A_3 : D_3)\}$,
  - $\tau_2 = \{(A_2 : D_2)\}$,

where both τ_1 and τ_2 are in $\mathfrak{T}_1$ and $\mathfrak{T}_2$, and $\mathfrak{T}_1 = \mathfrak{T}_2$.

□

4.1.4 Constructivity of Relational Backgrounds

By the term *relational background*, we refer to the background that is needed by computations over states described with relational models. In [29], two general purposes that a background may serve were discussed: (1) to describe available structures a computation may use and (2) to deal with new elements imported into states. These purposes are certainly valid for a relational background. Nevertheless, the notion of state thus far is too general to reflect the specific data environment of a database, for instance, database schemata, data integrity, etc. For this, we can restrict relational backgrounds via the provision of relational type schemes to define admissible states of a relational database transformation.

Definition 4.1.4. A *relational background* is a pair $(K^{rel}, \tilde{\Gamma}^{rel})$ consisting of a *relational background class* K^{rel} and a set $\tilde{\Gamma}^{rel}$ of *relational type schemes.*

In the following we continue to exploit the constructivity of relational backgrounds. By constructivity we mean the way that a background structure can be constructed from the atoms. Several notions of constructivity for backgrounds have been proposed by Blass and Gurevich [29]. Following their notions, we discuss the constructivity of relational backgrounds in terms of Relational Algebra (RA) since the algebraic properties of RA have been widely used for query optimisation in relational databases.

From weakest to strongest, we recall the notions of constuctivity defined by Blass and Gurevich [29]. The set $atom(S)$ of atoms in a state S consists of the elements in the base domains of S. The notation $atom(t)$ denotes the set of all atoms occurring in a term t.

Definition 4.1.5. A background class K is *rigid relative to atoms* if the only automorphism that fixes all the atoms is the identity function.

Definition 4.1.6. A background class K is *explicitly atom-generated* if the smallest substructure that contains all the atoms is K itself.

It is easy to see that each explicitly atom-generated background K is rigid relative to atoms. Moreover, all elements of an explicitly atom-generated background class are values of terms with atoms assigned to variables as values. Based on the notion of explicitly atom-generated, the notion of freely atom-generated is defined.

Definition 4.1.7. A background class K is *freely atom-generated* via a subset $\Upsilon_0 \subseteq \Upsilon$ and a set $\widehat{ID}$ of identities over Υ_0 if

- K is explicitly atom-generated, and
- two terms are equal (i.e., have the same value) in K if the equality can be derived from the identities in $\widehat{ID}$.

From an algebraic point of view, when considering functional symbols as algebraic operators, the set of identities would correspond to algebraic identities capturing equivalences between algebraic terms in a background class. Such a set of algebraic identities defines an equivalence relationship over terms in which two terms are equivalent if one term can be rewritten to the other based on these identities. Let $term(K)$ be a set of all terms of a background class K. Then this kind of equivalence relation is called *t-equivalent* (denoted as $\sim^t$: $term(K)^2 \rightarrow \{true, false\}$). On the other hand, given two terms t_1 and t_2 of a background class K, t_1 and t_2 are called *i-equivalent* (denoted as $\sim^i$: $term(K)^2 \rightarrow \{true, false\}$) if the interpretation of t_1 and t_2 yields the same value under the isomorphisms of K. Therefore, the set of algebraic identities is *sound* iff for all $t_1, t_2 \in term(K)$, $t_1 \sim^t t_2$ implies $t_1 \sim^i t_2$, and the set of algebraic identities is *complete* iff for all $t_1, t_2 \in term(K)$, $t_1 \sim^i t_2$ implies $t_1 \sim^t t_2$.

For database transformations, the existence of algebraic identities in a background class is not trivial. Indeed, a set of complete and sound algebraic identities provides a way to capture all semantically equivalent one-step transformations, which is useful for optimising the implementation of database transformations. In the following we take RA as an example to illustrate the presence of algebraic identities in most of relational backgrounds.

Relational Algebra (RA) is a data manipulation language in the RDM. It has six primitive operators defined on the basis of set theory. The operators are unary or binary (i.e. take as input one or two relations) but always result in a single output relation which may or not be over a different relation schema to those of the input relations. Let S be a state, R_E be a binary equality relation over atoms in S such that $R_E = \{(a, a) | a \in atom(S)\}$[1], relation symbols R_1, R_2 and R_3 be assigned with the same type $typ(R_1) = typ(R_2) = typ(R_3) = \{(A_1 : D_1, ..., A_n : D_n)\}$. Then the operators of RA are defined by

- *selection:*
  - $\sigma_{A_i=a}(R_1) = \{t | t \in I(R_1) \wedge t(A_i) = a\}$ or
  - $\sigma_{A_i=A_j}(R_1) = \{t | t \in I(R_1) \wedge t(A_i) = t(A_j)\}$,

  where $attr(\sigma_{A_i=a}(R_1)) = attr(\sigma_{A_i=A_j}(R_1)) = attr(R_1)$;

[1] R_E is used for testing the equality of two atoms.

- *projection*:
  - $\pi_{\mathcal{A}}(R_1) = \{t | \exists t' \in I(R_1) \forall A \in \mathcal{A}.t(A) = t'(A)\}$,

  where $attr(\pi_{\mathcal{A}}(R_1)) = \mathcal{A}$ for $\mathcal{A} \subseteq attr(R_1)$;

- *natural join*:
  - $R_1 \bowtie R_2 = \{t | \exists t_1 \in I(R_1), t_2 \in I(R_2) \forall A \in attr(R_1).t(A) = t_1(A) \wedge \forall A' \in attr(R_2).t(A') = t_2(A')\}$,

  where $attr(R_1 \bowtie R_2) = attr(R_1) \cup attr(R_2)$;

- *set union*:
  - $R_1 \cup R_2 = \{t | t \in I(R_1) \vee t \in I(R_2)\}$,

  where $attr(R_1 \cup R_2) = attr(R_1) = attr(R_2)$;

- *set different*:
  - $R_1 - R_2 = \{t | t \in I(R_1) \wedge t \notin I(R_2)\}$,

  where $attr(R_1 - R_2) = attr(R_1) = attr(R_2)$;

- *renaming*:
  - $\varrho_{A_i \mapsto A'}(R_1) = \{t | \exists t' \in I(R_1).t(A') = t'(A_i) \wedge \forall A \in attr(R_1) - \{A_i\}.t(A) = t'(A)\}$,

  where $attr(\varrho_{A_i \mapsto A'}(R_1)) = attr(R_1) - \{A_i\} \cup \{A'\}$.

These operators provide a means of transforming relational databases and thus need to be included in the relational background of RA. Furthermore, the background class of RA is explicitly atom-generated as all its elements can be obtained as closed terms with atoms assigned to variables as values. Since the second condition in Definition 4.1.7 is also satisfied, the background class of RA is freely atom-generated.

Let us consider elements of a relational background class for RA as terms. The set of terms of RA can be defined inductively by applying function symbols for its operators over relations at the database part of a state:

$$t \equiv R \mid \sigma_{A_i=a}(t) \mid \sigma_{A_i=A_j}(t) \mid \pi_{\mathcal{A}}(t) \mid t_1 \bowtie t_2 \mid t_1 \cup t_2 \mid t_1 - t_2 \mid \varrho_{A_i \to A'}(t).$$

In this sense, query expressions of RA are terms under compositions of algebraic operations. They can be optimised by rewriting query expressions in accordance with a set of relational algebraic identities. This is the rationale of query optimisation implemented in most of commercial relational databases. The following are some identities used in RA.

- *idempotent laws*:
  - $\sigma_\varphi(\sigma_\varphi(R)) = \sigma_\varphi(R)$
  - $\pi_{\mathcal{A}}(\pi_{\mathcal{A}}(R)) = \pi_{\mathcal{A}}(R)$
  - $R \bowtie R = R$
  - $R \cup R = R$
- *commutative laws*:
  - $\sigma_{\varphi_1}(\sigma_{\varphi_2}(R)) = \sigma_{\varphi_2}(\sigma_{\varphi_1}(R))$
  - $R_1 \bowtie R_2 = R_2 \bowtie R_1$
  - $R_1 \cup R_2 = R_2 \cup R_1$ provided $attr(R_1) = attr(R_2)$
- *associative laws*:
  - $(R_1 \bowtie R_2) \bowtie R_3 = R_1 \bowtie (R_2 \bowtie R_3)$
  - $(R_1 \cup R_2) \cup R_3 = R_1 \cup (R_2 \cup R_3)$ provided $attr(R_1) = attr(R_2) = attr(R_3)$
- *identity laws*:
  - $R \cup \emptyset = R$
  - $R - \emptyset = R$
- *conjunctive laws*:
  - $\sigma_{\varphi_1}(\sigma_{\varphi_2}(R)) = \sigma_{\varphi_2 \wedge \varphi_1}(R))$

- *other laws*:
  - $\pi_{\mathcal{A}_1}(\pi_{\mathcal{A}_2}(R)) = \pi_{\mathcal{A}_1}(R)$ (provided $\mathcal{A}_1 \subseteq \mathcal{A}_2$)
  - $\sigma_\varphi(\pi_{\mathcal{A}}(R)) = \pi_{\mathcal{A}}(\sigma_\varphi(R))$ (provided for every attribute A' occurring in φ $A' \in \mathcal{A}$)
  - ......

By adding an assignment rule $R := t$ for a relational symbol $R \in \Re_{out}$ and a term t, the query results of RA can be stored. As a well-known fact, a relational database management system uses more or less similar relational algebra operations for query execution and optimisation at an internal implementation level. Therefore, it is a common practice that relational algebra operations and identities are part of a relational background. Nevertheless, the expressive power of RA as a standalone database query language is still rather limited, for example, RA can not express the query *transitive closure* as no recursion mechanism is provided by RA.

4.1.5 Relational Database Transformations

Formally, we define relational database transformations as follows.

Definition 4.1.8. A *relational database transformation* is a database transformation with a relational background $(K^{rel}, \tilde{\Gamma}^{rel})$, satisfying the following conditions:

- the database part of the state signature contains a finite, non-empty set of relational symbols,
- the database part of each initial or final state is a relational database instance over an input or output relational database schema defined by the relational type scheme $\tilde{\Gamma}^{rel}$, respectively.

Intuitively speaking, a relational database transformation describes a process starting from a relational database instance constrained by an input relational schema, processing them in accordance with the postulates defined in Section 3.1 and terminating with a relational database instance constrained by an output relational schema. RA can be seen as one subclass of relational database transformations. By similarly formalising terms and backgrounds for other relational query languages, we arrive at different subclasses of relational database transformations.

We begin with *Query Language* (QL), which is a "complete" language in the sense of Chandra and Harel's definition of completeness proposed in [52]. QL concerns only the computations over finite relations residing at the database part of a state. Operations related to an infinite domain at the algorithmic part are disregarded. The terminology of QL has a procedural style like procedural programming languages. QL can essentially be regarded as being the closure of RA under certain programming primitives such as sequential composition and iteration with the ability to test for emptiness. Therefore, we present the syntax and semantics of QL by modifying and extending the definition of RA.

Let S be a state, $R \in \Re$ and $R_E \in \Re$ be a binary equality relation extending the one in RA such that $R_E = \{(a,a)|a \in atom(S) \cup \mathcal{P}(atom(S))\}^2$. Then the terms of QL are defined inductively in the same way as the terms of RA, such that

$$t \equiv R \mid \sigma_{A_i=a}(t) \mid \sigma_{A_i=A_j}(t) \mid \pi_{\mathcal{A}}(t) \mid t_1 \bowtie t_2 \mid t_1 \cup t_2 \mid t_1 - t_2 \mid \varrho_{A_i \to A'}(t).$$

The programs of QL extend the assignment rules of RA by adding sequential composition and while construct,

$$p \equiv R := t \mid p_1; p_2 \mid while \;\; R \;\; do \;\; p.$$

Semantically, $p_1; p_2$ denotes that p_1 and p_2 are executed sequentially, and $while \;\; R \;\; do \;\; p$ denotes that while R is an empty set then program p is executed. Thus, QL is equipped with the capability to deal with recursion by the use of while construct and counters simulated by operating on the ranks of relations and testing their emptiness. In doing so, the query *transitive closure* is expressible in QL. Despite all this, QL is still not expressive enough to deal with some common relational queries involving generalised aggregations, for instance, the query *counting* that counts the number of tuples in a relation satisfying certain conditions, or the query *summation* that sums up values of an attribute of relations under specified conditions.

The background of QL includes all the function symbols for relational algebra operators and the set of relational algebraic identities. Furthermore, it extends the computation background of RA by adding the extended equality relation and a function symbol for the test of empty relation in the background signature. The background class of QL is explicitly atom-generated.

[2]The extension of R_E over $\mathcal{P}(atom(S))$ is to provide the capability for testing the emptyness of a relation by utilising unranked relation variables. In doing so, counters can be simulated.

The observed limited expressiveness of QL motivated Chandra and Harel to propose an extension called Extended Query Language (EQL) in [52]. The intention is to take operations over potentially infinite domains into consideration. This naturally leads to consideration of a meta-finite structure where not only the database part but also the algorithmic part of a state play a role in computations.

Let S be a state with the signature $\Upsilon = \Upsilon_{db} \cup \Upsilon_a \cup \{f_1, ..., f_\ell\}$, $R \in \Re$, $R_E \in \Re$ and $R_E = \{(a,a) | a \in atom(S) \cup \bigcup \mathcal{P}(atom(S))\}$. Then the terms of EQL are defined by enlarging the terms of QL such that

(1) terms in the database part are closed under algebraic operators,

$$t \equiv R \mid \sigma_{A_i=a}(t) \mid \sigma_{A_i=A_j}(t) \mid \pi_{\mathcal{A}}(t) \mid t_1 \bowtie t_2 \mid t_1 \cup t_2 \mid t_1 - t_2 \mid \varrho_{A_i \to A'}(t).$$

(2) $f(t_1, ..., t_n)$ is a term in the algorithmic part for n-ary $f \in \{f_1, ..., f_\ell\}$ and $t_1, ..., t_n$ are terms in the database part, and

(3) $f(t_1, ..., t_n)$ is a term in the algorithmic part for n-ary $f \in \Upsilon_a$ and $t_1, ..., t_n$ are terms in the algorithmic part.

The programs of EQL are exactly the same as those of QL. They are,

$$p \equiv R := t \mid p_1; p_2 \mid while\ R\ do\ p.$$

With the additional terms obtained by applying bridge and algorithmic functions, EQL is expressive enough to capture generalised aggregations. For instance, the query *counting* can be expressed in EQL by defining a bridge function $card : D \to \mathbb{N}$ and an algorithmic function *sum* over $\mathbb{N}$ in the background. The background of EQL should include all the function symbols for relational algebra operators and the set of relational algebraic identities. However, different from the backgrounds of RA and QL, the background of EQL requires a type constructor for finite multiset as it is indispensable to capture the cardinality of elements occurring in a set. The inclusion of a type constructor for finite multiset and function symbols for its operations in a computation background serves as the basis for expressing generalised aggregation computing.

We use DB-ASM^{rel} to refer to the restricted class of DB-ASMs for relational database transformations. More precisely, a DB-ASM^{rel} is a DB-ASM in which

- the background is a relational background $(K^{rel}, \tilde{\Gamma}^{rel})$, and
- the database part of a state is a relational database instance.

Theorem 4.1.1. *DB-ASMs with relational backgrounds $(K^{rel}, \tilde{\Gamma}^{rel})$, in which initial and final states are relational database instances of an input and output relational database schema defined by the relational type scheme $\tilde{\Gamma}^{rel}$, respectively, capture exactly all relational database transformations with the same signature and relational background $(K^{rel}, \tilde{\Gamma}^{rel})$.*

Proof. This theorem is a direct consequence of Theorems 3.2.1 and 3.3.1. □

By taking the terms of EQL as the terms of DB-ASM^{rel}, DB-ASM^{rel} can express all the queries captured by EQL. The program constructs provided in EQL can be simulated by using DB-ASM^{rel} rules. More specifically, an assignment rule of DB-ASM^{rel} is more general than the one used in EQL since an assignment rule $R := t$ of DB-ASM^{rel} may have $R \in \Re_{in} \cup \Re_{out} \cup \Re_{temp}$. It means that an assignment rule of DB-ASM^{rel} can be used to update input relations in addition to enacting queries over them. A sequential rule of EQL can be simulated by a sequence rule of DB-ASM^{rel}. The power of the recursion provided by the while construct and the testing for emptiness in EQL can be supplied by using conditional and choice rules under recursively iterations in DB-ASM^{rel}. With the presence of choice rules of DB-ASM^{rel}, certain kinds of non-deterministic queries can also be expressed. Therefore, the following corollary is a straightforward result.

Corollary 4.1.1. RA $\subsetneq$ QL $\subsetneq$ EQL $\subsetneq$ DB-ASM^{rel}

4.2 Trees and Tree Algebra

As XML documents are often represented as trees, XML database transformations have to perform computations on trees. Thus tree values have to be available in the base set, and operations on such trees must be available for the definition of terms. Furthermore, tree values and operations have to be defined in the background. In this section, we provide all the necessary constituents of the backgrounds for XML database transformations, which leads to the definition of *tree background class* in Section 4.4.

4.2.1 Trees and Contexts

It is common to regard an XML document as an unranked tree, in which nodes may have an unbounded but finite number of children nodes.

Definition 4.2.1. An *unranked tree* γ is a structure $(\mathcal{O}_\gamma, \prec_c, \prec_s)$, consisting of

- a finite, non-empty set $\mathcal{O}_\gamma$ of node identifiers, called *tree domain*,
- ordering relations $\prec_c$ and $\prec_s$ over $\mathcal{O}_\gamma$ called *child relation* and *sibling relation*, respectively, satisfying the following conditions:
  - there exists a unique, distinguished node $o_r \in \mathcal{O}_\gamma$ (called the *root* of the tree) such that for all $o \in \mathcal{O}_\gamma - \{o_r\}$ there is exactly one $o' \in \mathcal{O}_\gamma$ with $o' \prec_c o$, and
  - whenever $o_1 \prec_s o_2$ holds, then there is some $o \in \mathcal{O}_\gamma$ with $o \prec_c o_i$ for $i = 1, 2$.

The relations $\prec_c$ and $\prec_s$ are irreflexive ($x \not\prec x$).

For $x_1 \prec_c x_2$ we say that x_2 is a *child* of x_1; for $x_1 \prec_s x_2$ we say that x_2 is the *next sibling* to the right of x_1.

In order to obtain XML trees from this, we require the nodes of an unranked tree to be labelled, and the leaves, i.e. nodes without children, to be associated with values. Let us fix a finite, non-empty set Σ of *labels*, and a finite family $\{\tau_i\}_{i \in I}$ of data types. Each data type τ_i is associated with a *value domain* D_{τ_i}. The corresponding *universe* U contains all possible values of these data types, i.e. $U = \bigcup_{i \in I} D_{\tau_i}$.

Definition 4.2.2. An *XML tree* t_γ (over the set Σ of labels with values in the universe U) is a triple $(\gamma, \omega, \upsilon)$ consisting of

- an unranked tree $\gamma = (\mathcal{O}_\gamma, \prec_c, \prec_s)$,
- a total *label function* ω: $\mathcal{O}_\gamma \to \Sigma$, and
- a partial *value function* υ: $\mathcal{O}_\gamma \to U$ such that
  - whenever υ is defined on o, o is a leaf in γ.

We use $root(t_\gamma)$ to denote the root node of an XML tree t_γ. Given two XML trees t_1 and t_2, t_1 is the *subtree* of t_2 if the following properties are satisfied: (1) $\mathcal{O}_{t_1} \subseteq \mathcal{O}_{t_2}$, (2) $o_1 \prec_c o_2$ holds in t_1 iff it holds in t_2, (3) $o_1 \prec_s o_2$ holds in t_1 iff it holds in t_2, (4) $\omega_{t_1}(o') = \omega_{t_2}(o')$ holds for all $o' \in \mathcal{O}_{t_1}$, and (5) either $\upsilon_{t_1}(o') = \upsilon_{t_2}(o')$ holds or otherwise both sides are undefined for all $o' \in \mathcal{O}_{t_1}$. t_1 is said to be the *largest subtree* of t_2 at node o, denoted as $\widehat{o}$, iff (1) t_1 is the subtree of t_2 with $root(t_1) = o$ and (2) there does not exist an

XML tree t_3 with $t_3 \neq t_1$ and $t_3 \neq t_2$ such that t_1 is the subtree of t_3 and t_3 is the subtree of t_2. The set of all XML trees over Σ – neglecting the universe U – is denoted as $\mathcal{T}(\Sigma)$.

A sequence $t_1, ..., t_k$ of XML trees is called an *XML hedge* or simply a *hedge*, and a multiset $\{\!\{t_1, ..., t_k\}\!\}$ of XML trees is called an *XML forest* or simply a *forest*. The notion of a forest is indispensable in situations where order is irrelevant, (e.g. when representing attributes of a node) or where duplicates are desirable (e.g. for computations in parallel on identical subtrees). Let ε denote the *empty hedge*.

As observed at the beginning of the chapter, in order to define flexible operations on XML trees, it is necessary to be able to select tree portions of interest. One such portion is subtree, but occasionally we will need more general structures. This will be supported by XML contexts.

Definition 4.2.3. An *XML context* t_c over the set of labels Σ ($\xi \notin \Sigma$) is an unranked tree over $\Sigma \cup \{\xi\}$ such that for each tree

- exactly one leaf node is labelled with the symbol ξ and has undefined value, and
- all other nodes in a tree are labelled and valued in the same way as an XML tree defined in Definition 4.2.2.

The context with a single node labelled by ξ is called the *trivial context* (also denoted as ξ). Similarly, the set of all XML contexts over Σ is denoted as $\mathcal{T}(\Sigma \cup \{\xi\})$.

We can now define substitution operations that replace a subtree of an XML tree or context by a new XML tree or context. When defining these substitutions, we have to ensure that the special label ξ occurs at most once in the result. For this reason, substitution from some XML context, whether it be by XML tree or context, will always be carried out for the node labelled ξ. This leads to the following distinction between four kinds of substitutions:

Tree-to-tree substitution

For an XML tree $t_{\gamma_1} \in \mathcal{T}(\Sigma_1)$ with a node $o \in \mathcal{O}_{\gamma_1}$ and an XML tree $t_{\gamma_2} \in \mathcal{T}(\Sigma_2)$, the result $t_{\gamma_1}[t_{\gamma_2}/\widehat{o}]$ of substituting t_{γ_2} for the subtree rooted at o is an XML tree in $\mathcal{T}(\Sigma_1 \cup \Sigma_2)$.

Tree-to-context substitution

For an XML tree $t_{\gamma_1} \in \mathcal{T}(\Sigma_1)$ with a node $o \in \mathcal{O}_{\gamma_1}$, the result $t_{\gamma_1}[\xi/\widehat{o}]$ of substituting the trivial context ξ for the subtree rooted at o is an XML context in $\mathcal{T}(\Sigma_1 \cup \{\xi\})$.

Context-to-context substitution

For an XML context $t_{c_1} \in \mathcal{T}(\Sigma_1 \cup \{\xi\})$ and an XML context $t_{c_2} \in \mathcal{T}(\Sigma_2 \cup \{\xi\})$, the result $t_{c_1}[t_{c2}/\xi]$ of substituting t_{c_2} for the node labelled ξ in t_{c_1} is an XML context in $\mathcal{T}(\Sigma_1 \cup \Sigma_2 \cup \{\xi\})$.

Context-to-tree substitution

For an XML context $t_{c_1} \in \mathcal{T}(\Sigma_1 \cup \{\xi\})$ and an XML tree $t_{\gamma_2} \in \mathcal{T}(\Sigma_2)$, the result $t_{c_1}[t_{\gamma_2}/\xi]$ of substituting t_{γ_2} for the node labelled by ξ in t_{c_1} is an XML tree in $\mathcal{T}(\Sigma_1 \cup \Sigma_2)$.

The correspondence between an XML document and an XML tree is straightforward. Each element of an XML document corresponds to a node of the XML tree and the subelements of an element define the children nodes of the node corresponding to the element. The nodes for elements are labeled by element names and character data of an XML document correspond to values of leaves in an XML tree. As our main focus is on structural properties of an XML document, attributes are handled as if they are subelements (i.e. they are not taken into particular consideration to simplify the discussion).

4.2.2 Tree Selector Constructs

To enable manipulation over XML trees at higher levels than individual nodes and edges, we need some tree constructs that are able to select arbitrary tree portions of interest. For this, we provide two selector constructs: *subtree* and *context*.

Definition 4.2.4. Let $t_\gamma = (\gamma, \omega, v)$ be an XML tree. Then the constructs *subtree* and *context* are defined as follows:

- *context* is a binary, partial function defined on pairs (o_1, o_2) of distinct nodes with $o_i \in \mathcal{O}_\gamma$ $(i = 1, 2)$ such that o_1 is an ancestor of o_2 (i.e. $o_1 \prec_c^* o_2$ holds for the transitive closure $\prec_c^*$ of $\prec_c$). We have

$$context(o_1, o_2) = \widehat{o}_1[\xi/\widehat{o}_2].$$

- *subtree* is a unary function defined on $\mathcal{O}_\gamma$. We have

$$subtree(o) = \widehat{o}.$$

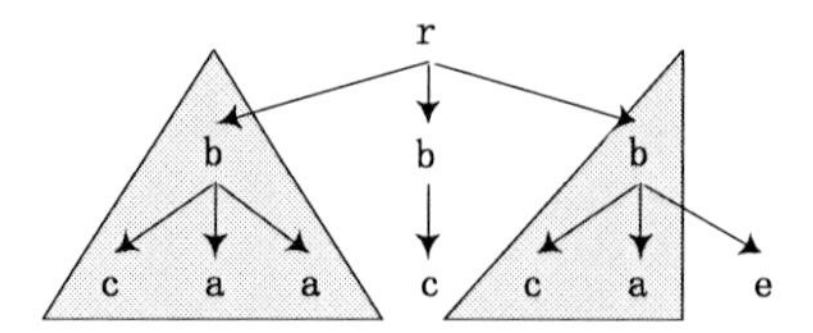

Figure 4.1: XML tree portions

Example 4.2.1. Consider the XML tree t_γ shown in Figure 4.1. Suppose that we want to select

1. a subtree, rooted at a node labelled by b, which has exactly two children nodes labelled by a, and

2. a context, defined by a subtree that is rooted at a node labelled by b, in which a subtree with a root also labelled by b is substituted by the trivial context ξ.

For (1) using DB-ASM rules we could build the following computational fragment:

forall x **with** $\exists x_1, x_2.((x \prec_c x_1) \wedge (x \prec_c x_2) \wedge \omega(x) = b \wedge \omega(x_1) = a \wedge \omega(x_2) = a \wedge$
$x_1 \neq x_2 \wedge \forall x_3((x \prec_c x_3) \wedge \omega(x_3) = a \Rightarrow (x_3 = x_1 \vee x_3 = x_2)))$

do

$t_1 := \mathit{subtree}(x)$; ...

enddo

Of course, the formula in the **with**-clause has to be interpreted using the XML tree t_γ with variables bound to identifiers in the tree domain $\mathcal{O}_\gamma$. Furthermore, $\prec_c$ has to be interpreted by the children relation, and ω by the labelling function. Such formulae and their interpretations will be discussed in Section 4.3.

Similarly, for (2) we could use the following fragment of a DB-ASM rule:

forall x **with** $(x \prec_c^* x_1) \wedge \omega(x) = b \wedge \omega(x_1) = b$

do

$t_2 := \mathit{context}(x, x_1)$; ...

enddo

The two resulting tree portions of interest are highlighted in Figure 4.1. The left tree portion corresponds to the subtree requested in (1), while the right tree portion corresponds to the context requested in (2). □

The use of the tree selector constructs $context(x_1, x_2)$ and $subtree(x)$, along with the tree algebra operations which we will introduce in the next subsection, allows us to extract and recombine portions of an existing XML tree to form new XML trees. This enables a flexible level of abstraction for XML database transformations beyond manipulation of nodes and edges.

4.2.3 Tree Algebra

To manipulate over tree structures, we need operators on forests or hedges. This motivates us to take into account tree algebras. These algebras have, however, been developed for many different purposes. For instance, the forest algebra by Mikolaj Bojanczyk and Igor Walukiewicz in [41] aims at providing an algebraic framework for studying logical definability of different classes of tree languages, while the tree algebra for binary ranked trees by Wilke in [157] was defined for characterising the class of frontier-testable tree languages satisfying conditions such as "there exists a natural number k such that any two trees with the same set of subtrees of depth at most k either belong both to the language or both not". In this subsection, we will develop an XML tree algebra that adapts features from these existing tree algebras towards the setting of XML trees. The main motivation of the XML tree algebra is to find an algebraic approach that can manipulate portions of a tree structure at a highly flexible abstraction level, such as subtrees and contexts. In doing so, individual nodes and edges are considered to be special cases of subtrees and contexts.

We now define a many-sorted algebra using three sorts: **L** for labels, **H** for hedges, and **C** for contexts, along with a set $\{f_\iota, f_\nu, f_\varsigma, f_\varrho, f_\kappa, f_\vartheta, f_\eta\}$ of function symbols with the following signatures:

- $f_\iota : \mathbf{L} \times \mathbf{H} \to \mathbf{H}$

- $f_\nu : \mathbf{L} \times \mathbf{C} \to \mathbf{C}$

- $f_\varsigma : \mathbf{H} \times \mathbf{C} \to \mathbf{C}$

- $f_\varrho : \mathbf{H} \times \mathbf{C} \to \mathbf{C}$

- $f_\kappa : \mathbf{H} \times \mathbf{H} \to \mathbf{H}$

- $f_\vartheta : \mathbf{C} \times \mathbf{C} \rightarrow \mathbf{C}$

- $f_\eta : \mathbf{C} \times \mathbf{H} \rightarrow \mathbf{H}$

Given a fixed set Σ of labels and two special symbols ε and ξ, the set $\mathcal{T}$ of terms over $\Sigma \cup \{\varepsilon, \xi\}$ comprises label terms, hedge terms, and context terms. That is, $\mathcal{T} = \mathcal{T}_L \cup \mathcal{T}_H \cup \mathcal{T}_C$, where $\mathcal{T}_L$, $\mathcal{T}_H$ and $\mathcal{T}_C$ stand for the sets of terms over sorts $\mathbf{L}$, $\mathbf{H}$ and $\mathbf{C}$, respectively.

- The set $\mathcal{T}_L$ of label terms is simply the set of labels, i.e. $\mathcal{T}_L = \Sigma$.
- The set $\mathcal{T}_H$ of hedge terms is defined by:
  - $\mathcal{T}_H^s \subseteq \mathcal{T}_H$ where $\mathcal{T}_H^s$ is the set of tree terms corresponding to XML trees,
  - $\varepsilon \in \mathcal{T}_H^s$,
  - $t\langle t' \rangle \in \mathcal{T}_H^s$ for $t \in \Sigma$ and $t' \in \mathcal{T}_H$, and
  - $t_1, ..., t_n \in \mathcal{T}_H$ for $t_i \in \mathcal{T}_H^s$ $(i = 1, ..., n)$.
- The set of context terms $\mathcal{T}_C$ is the smallest set with
  - $\xi \in \mathcal{T}_C$, and
  - $t\langle t_1, ..., t_n \rangle \in \mathcal{T}_C$ for a label $t \in \Sigma$ and terms $t_1, ..., t_n \in \mathcal{T}_H^s \cup \mathcal{T}_C$ such that exactly one t_i $(i = 1, \ldots, n)$ is a context term in $\mathcal{T}_C$.

Correspondingly, tree, hedge and context terms represent XML trees, hedges and contexts that were defined earlier. In accordance with the first condition in the definition of hedge terms, XML trees can be identified with hedges of length 1. Moreover, the last condition in the definition of context terms guarantees that all context terms contain exactly one ξ. Trees and contexts have a root, but hedges do not (unless they can be identified with a tree). For hedges of the form $t\langle \varepsilon \rangle$, we use t as a notational shortcut, if there is no confusion with the label term t. Furthermore, we use $\#t$ to denote the sort of term t.

Example 4.2.2. Let $\Sigma = \{a, b, c\}$. Then we have

- a, b and c are terms of sort $\mathbf{L}$,
- $a\langle b\langle b \rangle, c\langle a \rangle \rangle$ and $b\langle c \rangle, a\langle a\langle b \rangle, b \rangle$ are terms of sort $\mathbf{H}$, and
- $a\langle a\langle b \rangle, \xi, c \rangle$ and $a\langle a\langle b, \xi \rangle, c \rangle$ are terms of sort $\mathbf{C}$. □

Intuitively speaking, the functions f_ι and f_ν extend hedges and contexts upwards with labels, and f_ς and f_ϱ incorporate hedges into non-trivial contexts from left or right, respectively, which takes care of the order of subtrees arising in XML (see Example 4.2.3 to follow). The function f_κ denotes hedge juxtaposition, and likewise f_ϑ denotes context composition. The function f_η denotes context substitution, i.e. substituting the variable ξ in a context with a hedge, which leads to a tree.

Figure 4.2, in which tree terms are represented by triangles and context terms by triangles with a dot for ξ, provides further illustration of these functions. The following is their formal definitions:

$$f_\iota(t,(t_1,\dots,t_n)) = t\langle t_1,\dots,t_n\rangle \tag{4.1}$$

(The case $n=0$ leads to $t\langle\varepsilon\rangle$ on the right hand side.)

$$f_\nu(t,t^{'}) = t\langle t^{'}\rangle \tag{4.2}$$

$$f_\varsigma((t_1,\dots,t_n),t\langle t_1^{'},\dots,t_m^{'}\rangle) = t\langle t_1,\dots,t_n,t_1^{'},\dots,t_m^{'}\rangle \tag{4.3}$$

$$f_\varrho((t_1,\dots,t_n),t\langle t_1^{'},\dots,t_m^{'}\rangle) = t\langle t_1^{'},\dots,t_m^{'},t_1,\dots,t_n\rangle \tag{4.4}$$

$$f_\kappa((t_1,\dots,t_n),(t^1,\dots,t^m)) = t_1,\dots,t_n,t^1,\dots,t^m \tag{4.5}$$

$$f_\vartheta(t,t^{'}) = t[t^{'}/\xi] \tag{4.6}$$

$$f_\eta(t,(t_1,\dots,t_n)) = t[t_1,\dots,t_n/\xi] \tag{4.7}$$

Example 4.2.3. Let us have a look at Figure 4.3. Given a context term $t_2 = a\langle b,\xi\rangle$ and a hedge term $t_1 = b\langle c\rangle, b\langle c\rangle$, we obtain

- the context in (i) by $f_\varsigma(t_1,t_2) = a\langle b\langle c\rangle, b\langle c\rangle, b, \xi\rangle$, and
- the context in (ii) by $f_\varrho(t_1,t_2) = a\langle b,\xi,b\langle c\rangle,b\langle c\rangle\rangle$. □

We can now investigate equivalences among terms. The proof of the following proposition is straightforward.

Proposition 4.2.1. *The XML tree algebra satisfies the following equations for terms $t_1,t_2,t_3 \in \mathcal{T}$ of appropriate sorts (i.e., whenever one of the terms in the equation is defined,*

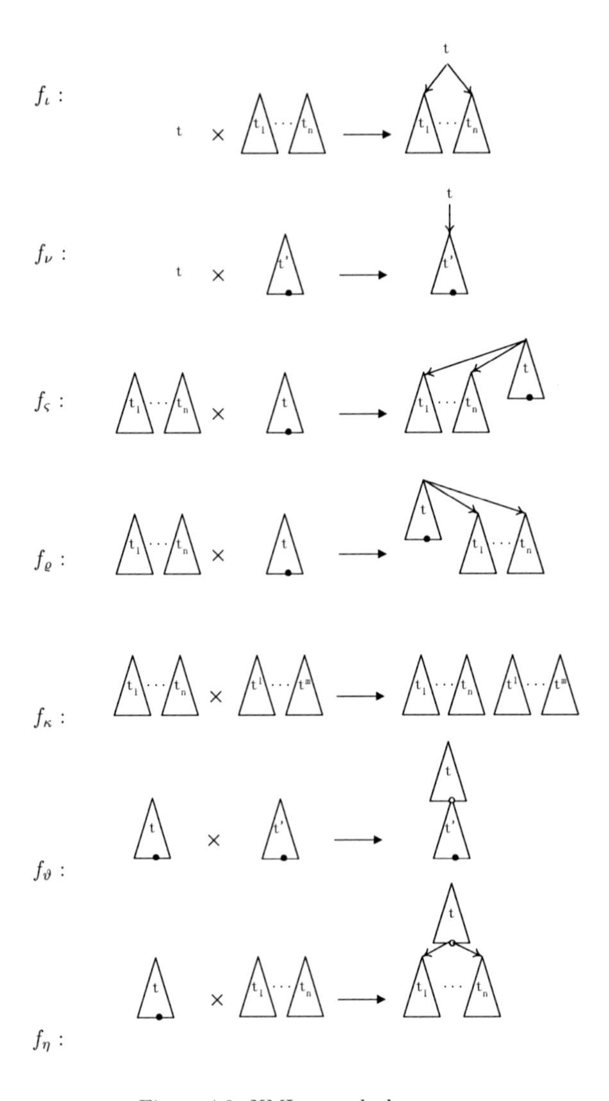

Figure 4.2: XML tree algebra

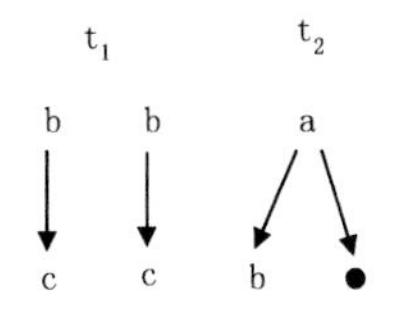

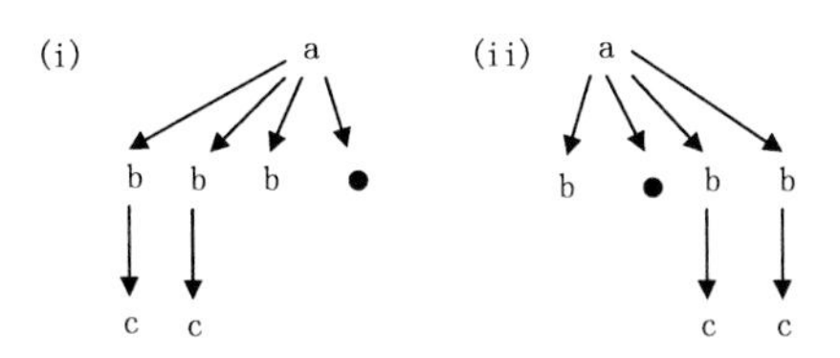

Figure 4.3: An illustration of functions $f_\varsigma(t_1, t_2)$ and $f_\varrho(t_1, t_2)$

the other one is defined, too, and equality holds):

$$f_\eta(f_\vartheta(t_1, t_2), t_3) = f_\eta(t_1, f_\eta(t_2, t_3)) \tag{4.8}$$

$$f_\vartheta(f_\vartheta(t_1, t_2), t_3) = f_\vartheta(t_1, f_\vartheta(t_2, t_3)) \tag{4.9}$$

$$f_\kappa(f_\kappa(t_1, t_2), t_3) = f_\kappa(t_1, f_\kappa(t_2, t_3)) \tag{4.10}$$

$$f_\eta(f_\nu(t_1, t_2), t_3) = f_\iota(t_1, f_\eta(t_2, t_3)) \tag{4.11}$$

$$f_\varsigma(t_1, f_\varsigma(t_2, t_3)) = f_\varsigma(f_\kappa(t_1, t_2), t_3) \tag{4.12}$$

$$f_\varrho(t_1, f_\varrho(t_2, t_3)) = f_\varrho(f_\kappa(t_2, t_1), t_3) \tag{4.13}$$

$$f_\varrho(t_3, f_\varsigma(t_1, t_2)) = f_\varsigma(t_1, f_\varrho(t_3, t_2)) \tag{4.14}$$

Example 4.2.4. To illustrate the correctness of equation (4.11) (i.e., $f_\eta(f_\nu(t_1, t_2), t_3) = f_\iota(t_1, f_\eta(t_2, t_3))$), let us take $t_1 = b$, $t_2 = a\langle b, \xi\rangle$ and $t_3 = b\langle c\rangle$. Then

- the left hand side becomes

$$f_\eta(f_\nu(t_1, t_2), t_3) = f_\eta(f_\nu(b, a\langle b, \xi\rangle), b\langle c\rangle) = f_\eta(b\langle a\langle b, \xi\rangle\rangle, b\langle c\rangle) = b\langle a\langle b, b\langle c\rangle\rangle\rangle$$

- the right hand side becomes

$$f_\iota(t_1, f_\eta(t_2, t_3)) = f_\iota(b, f_\eta(a\langle b, \xi\rangle, b\langle c\rangle)) = f_\iota(b, a\langle b, b\langle c\rangle\rangle) = b\langle a\langle b, b\langle c\rangle\rangle\rangle$$

□

Remark 4.2.1. In terms of manipulating tree structures, our tree algebra is powerful enough to express operations provided by the forest algebra [41] and Wilke's tree algebra [157]. For example, the operations $\iota^A(a)$, $\lambda^A(a,t)$ and $\rho^A(a,t)$ in [157] can be expressed as $\iota(a,\varepsilon)$, $\rho(\delta(a,\xi))$ and $\varsigma(\delta(a,\xi))$, respectively, in our tree algebra.

4.3 Weak Monadic Second-Order Logic

In Example 4.2.1, we have already seen formulae that can give rise to node identifiers, which then in turn can be used to construct tree portions of interest. In a sense, this illustrates the use of logic for navigating within an XML tree to identify tree portions of interest. Such navigation capability when used in conjunction with XML tree algebra enables us to specify manipulation over tree structures. In this section, we concentrate on the logical part.

We begin with providing a weak MSO logic over finite and unranked trees adopting from [61] in which second-order variables can be quantified only over finite sets. There are several reasons for choosing such a logic in our study.

Firstly, the use of MSO logic is motivated by its close correspondence to regular languages, which is known since early work of Büchi. Büchi [48] observed that a set of strings is regular iff it is definable in weak MSO logic with one successor. After that, Doner [65], Thatcher and Wright [140] extended this result to the case of trees. That is, a set of trees is regular iff it is definable in weak MSO logic with k successors. Nowadays, MSO logic has become important for XML database theory. The paper [61] provides a good introduction to these results.

Furthermore, we choose the weak version of MSO logic because DB-ASMs are restricted to use database variables in forall and choice rules, which must be interpreted by values in the database part of a state. Due to the abstract state postulate (i.e., Postulate 3.1.2), a finiteness condition is stipulated on the database part of any state of database transformations.

In most XML query and transformation languages, XPath [24] plays a key role in navigating tree structures. The navigational core of XPath2.0 captures the first-order logic. Some extensions on XPath, by adding transitive closure operators or using fixed-point operators, have been shown to be expressively complete for MSO logic or fragments of MSO logic [39, 138, 139]. For example, an extension called "regular XPath" with a Kleene star operator for transitive closure is complete with respect to the first-order logic extended with monadic transitive closure, which is strictly less expressive than MSO logic (over XML trees) [139, 139]. As we plan to study the effects of increasing the expressiveness of the logic incorporated in DB-ASM rules on the expressive power of the resulting computation model, MSO logic as an expressive logic over tree structures turns out to be a good candidate.

4.3.1 Tree Formulae

Let $\mathcal{X}_{FO}$ denote the set of first-variables expressed by lower-case letters and $\mathcal{X}_{SO}$ denote the set of second-order variables expressed by upper-case letters. Using abstract syntax, the formulae of MSO_X are defined by

$$\begin{aligned} \varphi \equiv\ & x_1 = x_2 \mid v(x_1) = v(x_2) \mid \omega(x_1) = a \mid x \in X \mid x_1 \prec_c x_2 \mid x_1 \prec_s x_2 \mid \\ & \neg\varphi \mid \varphi_1 \wedge \varphi_2 \mid \exists x.\varphi \mid \exists X.\varphi \end{aligned} \tag{4.15}$$

with $x, x_1, x_2 \in \mathcal{X}_{FO}$, $X \in \mathcal{X}_{SO}$, unary function symbols v and ω, $a \in \Sigma$ and binary predicate symbols $\prec_c$ and $\prec_s$.

We interpret formulae of MSO_X for a given XML tree $t_\gamma = (\gamma, \omega_t, v_t)$ over the set Σ of labels with $\gamma = (\mathcal{O}^t_\gamma, \prec^t_c, \prec^t_s)$. Naturally, the function symbols ω and v should be interpreted by the labelling and value functions ω_t and v_t, respectively, and the predicate symbols $\prec_c$ and $\prec_s$ should have interpretations in accordance with the children and sibling relations $\prec^t_c$ and $\prec^t_s$, respectively.

Furthermore, we need a variable assignment $\zeta : \mathcal{X}_{FO} \cup \mathcal{X}_{SO} \rightarrow \mathcal{O}_\gamma \cup \mathcal{P}(\mathcal{O}_\gamma)$ mapping first-order variables x to node identifiers $\zeta(x) \in \mathcal{O}_\gamma$, and second-order variables X to sets of node identifiers $\zeta(X) \subseteq \mathcal{P}(\mathcal{O}_\gamma)$. As usual, $\zeta[x \mapsto o]$ (and $\zeta[X \mapsto O]$, respectively) denotes the modified variable assignment, which equals ζ on all variables except the first-order variable x (or the second-order variable X, respectively), for which we have $\zeta[x \mapsto o](x) = o$ (and $\zeta[X \mapsto O](X) = O$, respectively).

With the XML tree t_γ and a variable assignment ζ, we obtain the interpretation $val_{t_\gamma,\zeta}$ on terms and formulae as follows. Terms in the logic have the form x, X or $v(x)$ where x is a first-order variable, X is a second-order variable and $v(x)$ is the value of some node associated with a first-order variable x, such that they are interpreted as

- $val_{t_\gamma,\varsigma}(x) = \varsigma(x)$,
- $val_{t_\gamma,\varsigma}(X) = \varsigma(X)$, and
- $val_{t_\gamma,\varsigma}(v(x)) = v_t(\varsigma(x))$.

For formula φ, we use $[\![\varphi]\!]_{t_\gamma,\varsigma}$ to denote its interpretation by a truth value and obtain:

- $[\![t_1 = t_2]\!]_{t_\gamma,\varsigma} = true$ iff $val_{t_\gamma,\varsigma}(t_1) = val_{t_\gamma,\varsigma}(t_2)$ holds for the terms t_1 and t_2,
- $[\![\omega(x) = a]\!]_{t_\gamma,\varsigma} = true$ holds iff $\omega_t(val_{t_\gamma,\varsigma}(x)) = a$,
- $[\![x \in X]\!]_{t_\gamma,\varsigma} = true$ iff $val_{t_\gamma,\varsigma}(x) \in val_{t_\gamma,\varsigma}(X)$,
- $[\![\neg\varphi]\!]_{t_\gamma,\varsigma} = true$ iff $[\![\varphi]\!]_{t_\gamma,\varsigma} = false$,
- $[\![\varphi_1 \wedge \varphi_2]\!]_{t_\gamma,\varsigma} = true$ iff $[\![\varphi_1]\!]_{t_\gamma,\varsigma} = true$ and $[\![\varphi_2]\!]_{t_\gamma,\varsigma} = true$,
- $[\![\exists x.\varphi]\!]_{t_\gamma,\varsigma} = true$ iff $[\![\varphi]\!]_{t_\gamma,\varsigma[x\mapsto o]} = true$ holds for some $o \in \mathcal{O}_\gamma$,
- $[\![\exists X.\varphi]\!]_{t_\gamma,\varsigma} = true$ iff $[\![\varphi]\!]_{t_\gamma,\varsigma[X\mapsto O]} = true$ for some finite $O \subseteq \mathcal{O}_\gamma$,
- $[\![x_1 \prec_c x_2]\!]_{t_\gamma,\varsigma} = true$ iff $val_{t_\gamma,\varsigma}(x_2)$ is a child node of $val_{t_\gamma,\varsigma}(x_1)$ in t_γ, i.e. $val_{t_\gamma,\varsigma}(x_1) \prec_c^t val_{t_\gamma,\varsigma}(x_2)$ holds, and
- $[\![x_1 \prec_s x_2]\!]_{t_\gamma,\varsigma} = true$ iff $val_{t_\gamma,\varsigma}(x_2)$ is the next sibling to the right of $val_{t_\gamma,\varsigma}(x_1)$ in t_γ, i.e. $val_{t_\gamma,\varsigma}(x_1) \prec_s^t val_{t_\gamma,\varsigma}(x_2)$ holds.

The syntax of MSO_X can be enriched by adding $\varphi_1 \vee \varphi_2$, $\forall x.\varphi$, $\forall X.\varphi$, $\varphi_1 \Rightarrow \varphi_2$, $\varphi_1 \Leftrightarrow \varphi_2$ as abbreviations for other MSO_X formulae in the usual way. Likewise, the definition of bound and free variables of MSO_X formulae is also standard. We use the notation $fr(\varphi)$ for the set of free variables in formula φ.

Definition 4.3.1. Given an XML tree t_γ and a MSO_X formula φ with $fr(\varphi) = \{x_1, \ldots, x_n\}$, then φ is said to be *satisfiable* in t_γ with respect to the variable assignment ς if $[\![\varphi]\!]_{t_\gamma,\varsigma} = true$.

4.3.2 Formulae in DB-ASMs with Trees

Straightforwardly, we can incorporate the logic MSO_X into the framework of DB-ASMs. When using XML trees as values in the base set of a DB-ASM, the logic MSO_X is not sufficient, as we have to take more than one such tree into account. Fortunately, this only requires an extension for the *atomic formulae*, i.e., those formulae appearing in the first line of Equation (4.15) above.

Because of their intimate connection to the definition of XML trees, the specific function and predicate symbols for XML trees, (i.e. υ, ω, $\prec_c$ and $\prec_s$) need to be extended by adding an XML tree as an additional argument. Thus, we obtain function symbols $\upsilon^{'}$, $\omega^{'}$, $\prec_c^{'}$ and $\prec_s^{'}$ of arity 2, 2, 3 and 3, respectively. That is,

- $\upsilon^{'}(x_1, x_2)$ denotes the value at leaf node x_2 in the tree x_1,
- $\omega^{'}(x_1, x_2)$ denotes the label of node x_2 in the tree x_1,
- $x_1 \prec_c^{'} (x) x_2$ denotes the (truth) value of $x_1 \prec_c x_2$ in the tree x, and
- $x_1 \prec_s^{'} (x) x_2$ denotes the (truth) value of $x_1 \prec_s x_2$ in the tree x.

Together with any other function symbols that are defined as part of the background signature and variables, we can build the set of terms. For this, variables now have to be sorted including sorts for node identifiers and sets of node identifiers, plus sorts for labels, hedges, contexts, etc.

- The set of atomic formulae then consists of formulae of the form $t_1 = t_2$, $x \in X$, $x_1 \prec_c^{'} (x^{'}) x_2$ and $x_1 \prec_s^{'} (x^{'}) x_2$ with
  - terms t_i $(i = 1, 2)$ of the same sort,
  - x, x_i $(i = 1, 2)$ of sort "node identifier",
  - X of sort " set of node identifier", and
  - $x^{'}$ of sort tree.
- The set of formulae is then built upon atomic formulae in the usual way using negation, conjunction and existential quantification plus the usual shortcuts as mentioned previously.

4.4 XML Database Transformations

In this section, we develop the general notion of XML database transformation. As outlined in Chapter 3, the customisation for a particular data model requires the definition of adequate backgrounds. We approach this in two steps. Firstly, we define tree background classes which build upon the XML tree algebra and weak MSO logic of the previous two sections. Then, we add tree type schemes to capture schema information which is based on EDTDs [116]. Thus, any object satisfying the postulates defined in Postulates 3.1.1, 3.1.2, 3.1.3, 3.1.4 and 3.1.5 with tree-based backgrounds (as defined in this section) is deemed to be an *XML database transformation.*

Using the same tree-based backgrounds, we obtain a computational model for XML database transformations on the basis of DB-ASMs, which were introduced in Chapter 3. From Theorems 3.2.1 and 3.3.1, we obtain that DB-ASMs with tree-based backgrounds capture exactly all XML database transformations.

4.4.1 Tree Background Classes

The background for an XML database transformation describes all available tree structures that may be used within an XML database transformation. Such backgrounds will be called *tree-based backgrounds.* In order to define tree-based backgrounds, a background signature (i.e. constructor symbols and function symbols) is first required.

Definition 4.4.1. A *tree-based background signature* Υ_K^{tree} contains at least

- a finite tuple constructor $(\cdot)$, a finite set constructor $\{\cdot\}$, a finite hedge constructor $, \dots ,$ and a finite tree constructor $\langle \cdot \rangle$,
- function symbols $root$, $\prec_c, \prec_s, \omega$ and v to define XML trees,
- function symbols *context* and *subtree* for the selection of tree portions,
- function symbols ε (empty hedge), ξ (trivial context),
- function symbols f_ι, f_ν, f_ς, f_ϱ, f_κ, f_η and f_ϑ as defined by the XML tree algebra,
- function symbols $\emptyset$ (empty set), $\{x\}$ (singleton set), $\cup$ (set union), and $\in$ (set membership) to deal with sets of node identifiers, and

- constructor symbols and function symbols as requested in the background postulate, i.e., Postulate 3.1.3.

In the sense of [47], all function symbols in a background signature are *static*, i.e. their interpretation is fixed and does not permit updates. On the other hand, the interpretation of functions symbols defined in the signature of underlying states can be updated.

Next we need a universe of elements, for which a set $\mathcal{D}$ of *base domains* is required analogous to Definition 3.1.7. Let us fix a set Σ of labels and a tree domain $\mathcal{O}$ (i.e. a set of node identifiers). Then Σ defines one of the base domains and $\mathcal{O}$ defines a *tree universe* as well as a set of *contexts*.

Definition 4.4.2. The *tree universe* over $\mathcal{O}$ is the set $HFT(\mathcal{O})$ of *hereditarily finite trees* over $\mathcal{O}$ which is the smallest set such that

- if $t_1, \dots, t_n \in \mathcal{O} \cup HFT(\mathcal{O})$, then $t\langle t_1, \dots, t_n\rangle \in HFT(\mathcal{O})$ for $t \in \mathcal{O}$.

For $n = 0$ in this definition we obtain trivial trees $t\langle\varepsilon\rangle \in HFT(\mathcal{O})$. If we identify $t\langle\varepsilon\rangle$ with t, we have in fact $\mathcal{O} \subseteq HFT(\mathcal{O})$. Each finite tree in a tree universe is indeed a tree skeleton in the sense that none of the nodes are labelled or assigned with values. By viewing hereditarily finite trees in $HFT(\mathcal{O})$ as special kinds of hereditarily finite lists, which can be interpreted as hereditarily finite sets by the Kuratowski encoding [150], the tree universe can also be treated as a special case of hereditarily finite sets.

In the spirit of Definition 4.4.2 we can define a set of *contexts* over $\mathcal{O}$ as follows.

Definition 4.4.3. The *set of contexts* over $\mathcal{O}$ is the smallest set $HFT(\mathcal{O}, \xi)$ with

- $\xi \in HFT(\mathcal{O}, \xi)$ and
- $t\langle t_1, \dots, t_n\rangle \in HFT(\mathcal{O}, \xi)$, where
  - $t \in \mathcal{O}$,
  - exactly one $t_i \in HFT(\mathcal{O}, \xi)$ for $i \in [1, n]$ is a context, and
  - all others $t_j \in HFT(\mathcal{O})$ for $j = 1, ..., n$ and $j \neq i$ are hereditarily finite trees.

In addition to the set Σ of labels, the tree universe $HFT(\mathcal{O})$ and the set of contexts $HFT(\mathcal{O}, \xi)$ define two more base domains.

Definition 4.4.4. A *tree background class* K^{tree} consists of

- a background signature Υ_K^{tree},
- a set $\mathcal{D}$ of base domains with $\Sigma \in \mathcal{D}, HFT(\mathcal{O}) \in \mathcal{D}, HFT(\mathcal{O}, \xi) \in \mathcal{D}$, and
- a *tree background structure* over Υ_K^{tree} and $\mathcal{D}$, which is a structure consisting of a universe U as defined in Definition 3.1.7 and an interpretation of function symbols in Υ_K^{tree} over U.

4.4.2 XML Schemata

XML documents may be associated with a schema formalism, for example, Document Type Definitions (DTDs), XML Schema, Extended Document Type Definitions (EDTDs) and so forth. Following the discussion in [111], we concentrate on EDTDs, as they generalise several popular formalisms. Research on EDTDs have also investigated how streaming XML documents can be validated using visibly pushdown automata or ASMs [125, 126, 129]. Adopting the discussion here to other XML schema formalisms is straightforward.

Given a set Σ of labels, the set of regular languages over Σ is denoted as $reg(\Sigma)$. We first define DTDs in an abstract way following [116]. In particular, we abstract from specific syntax of opening and closing tags and blur the distinction between subelements and attributes.

Definition 4.4.5. A *document type definition* (DTD) consists of

- a finite, non-empty set Σ of labels,
- a root $a_r \in \Sigma$, and
- a mapping $\beta : \Sigma \rightarrow reg(\Sigma)$ assigning to each $a \in \Sigma$ a regular language over Σ.

Note that this is called *labelled ordered tree object type definition* in [116].

Example 4.4.1. (adapted from [116]) Consider the following DTD:

⟨!DOCTYPE dealer [
  ⟨!ELEMENT dealer (used_cars, new_cars)⟩
  ⟨!ELEMENT used_cars (ad*)⟩
  ⟨!ELEMENT new_cars (ad*)⟩
  ⟨!ELEMENT ad ((model, year)|(model))⟩
  ⟨!ELEMENT model (#PCDATA)⟩

⟨!ELEMENT year (#PCDATA)⟩

]⟩

Using Definition 4.4.5, we can represent this DTD by the set of labels $\Sigma = \{\text{root}, \text{dealer}, \text{used\_cars}, \text{new\_cars}, \text{ad}, \text{model}, \text{year}\}$ with root 'root' and associated regular languages

- root: dealer

- dealer: used_cars new_cars

- used_cars: ad*

- new_cars: ad*

- ad: model year?

The assigned languages are

$$\begin{aligned}
\beta(\text{root}) &= \beta(\text{dealer}) & \beta(\text{dealer}) &= \text{dealer}\ \beta(\text{used_cars})\ \beta(\text{new_cars}) \\
\beta(\text{used_cars}) &= \text{used_cars}\ \beta(\text{ad})^* & \beta(\text{ad}) &= \text{ad}\ \beta(\text{model})\ (\beta(\text{year}) \cup \{\epsilon\}) \\
\beta(\text{new_cars}) &= \text{new_cars}\ \beta(\text{ad})^* & \beta(\text{model}) &= \{\text{model}\} \\
&\text{and} & \beta(\text{year}) &= \{\text{year}\}
\end{aligned}$$

□

The simplicity of DTDs has contributed greatly to its popularity among practitioners. However, it is also well-known that DTD is highly restricted in terms of what it can express. For instance, the DTD in Example 4.4.1 can not state that the 'year' tag must be present only for used cars. This could be avoided by having two different tags such as 'ad_used' and 'ad_new'. Extending DTD to Extended Document Type Definition (introduced in [116] as *specialised labelled ordered tree object type definition*) takes care of this problem.

Definition 4.4.6. An *extended document type definition* (EDTD) consists of

- a DTD (Σ', a_r, β), and

- a mapping $\alpha : \Sigma' \to \Sigma$ with another finite, non-empty set Σ of labels.

We can use the elements in Σ' to fine-tune the desired structure of XML document adhering to a given EDTD, while $\alpha(a)$ defines the actual tag that is to be used. In our example above, we could use 'ad_used' and 'ad_new' as elements of Σ' with both being mapped by α to 'ad' in Σ – all other elements of Σ' would be mapped to themselves.

Therefore, α captures specialisations between elements. Abstractly, we are not concerned with the particular name of a specialised element. Instead, we are more interested in knowing which element it is a specialisation of. Accordingly, we adopt the notational convention to write a^b for elements in Σ' where $\alpha(a^b) = a$. Normally, the superscript b of a^b is then called the *type* of the element, but to simplify the development in the next subsection we refer to $a^b \in \Sigma'$ as the type. If $\alpha^{-1}(a)$ contains only one element, we omit the superscript and assume that α maps a to itself.

Example 4.4.2. (adapted from [116]) The following denotes an EDTD with $\Sigma = \{\text{root}, \text{dealer}, \text{used\_cars}, \text{new\_cars}, \text{ad}, \text{model}, \text{year}\}$:

- root: dealer

- dealer: used_cars new_cars

- used_cars: $(\text{ad}^u)^*$

- new_cars: $(\text{ad}^n)^*$

- ad^u: model year

- ad^n: model

In any XML document adhering to this EDTD we would indeed have 'model' and 'year' for each used car, but only 'model' for new cars. □

Definition 4.4.7. Let t_γ be an XML tree over Σ', $G_1 = (\Sigma', a_r, \beta)$ be a DTD and $G_2 = (\Sigma, G_1, \alpha)$ be an EDTD. Then,

- t_γ is said to *satisfy* G_1 if the root of t_γ is labelled as a_r, and $a_1 \dots a_n \in \alpha(o)$ for each $o \in \mathcal{O}_\gamma$ labelled by a with children nodes labelled by $a_1, \dots, a_n$.

- The tree t_γ is said to *satisfy* G_2 if there exists an XML tree t'_γ over Σ' satisfying G_1 such that $\alpha(t'_\gamma) = t_\gamma$.

Here, we applied β to a whole XML tree, which must be understood as the canonical extension from node labels to trees.

We denote the set of XML trees satisfying a DTD or EDTD G as $sat(G)$, which can be considered as a regular language over Σ. Nevertheless, the set $\{sat(G) \mid G \text{ is a DTD over } \Sigma\}$ does not capture all the regular languages over Σ. This, however, holds for all XML trees satisfying an EDTD over Σ by straightforwardly applying the results in [116]:

Theorem 4.4.1. *(Papakonstantinou and Vianu [116]) A set of XML trees equals $sat(G)$ for some EDTD G over Σ iff it is a regular tree language over Σ.*

This result further justifies our choice to base the discussion of schema formalisms for XML database transformations on EDTDs.

4.4.3 Tree Type Schemes

As seen in the previous subsection, the labels of nodes in an XML tree are insufficient to express typing. Instead, we have to provide *types* in addition. In order to capture EDTDs we will define such types by type names (in Σ') and an association with regular expressions that are closed under union ($\cup$), concatenation ($|$) and Kleene star ($*$). For an EDTD $G_2 = (\Sigma, G_1, \alpha)$ with $G_1 = (\Sigma', a_r, \beta)$, the set of *type names* associated with XML trees in $sat(G_2)$ is Σ'. Furthermore, each type name $a \in \Sigma'$ is associated with a regular expression over Σ'.

Definition 4.4.8. Let $\mathcal{D}$ be a set of base domains with $\Sigma \in \mathcal{D}$, $HFT(\mathcal{O}) \in \mathcal{D}$ and $HFT(\mathcal{O}, \xi) \in \mathcal{D}$. Let $G_2 = (\Sigma, G_1, \alpha)$ be an EDTD with $G_1 = (\Sigma', a_r, \beta)$. Then a *tree type scheme* over $\mathcal{D}$ with respect to G_2 is a triple (Σ', typ_n, typ_e) consisting of

- a finite, non-empty set Σ' of type names,
- a *type name assignment* $typ_n : \mathcal{O} \to \Sigma'$ that associates with each node $o \in \mathcal{O}$ a type name $typ_n(o)$ in Σ', and
- a *type expression assignment* $typ_e : \Sigma' \to reg(\Sigma')$ that associates with each type name $a \in \Sigma'$ a regular expression $typ_e(a)$ over Σ'.

We write $a(\tau)$ to denote a type name $a \in \Sigma'$ together with its type expression $\tau = typ_e(a)$. $a(\tau)$ is interpreted by the set of trees $\{t\langle t' \rangle \mid t \in \mathcal{O}, \omega(t) = \alpha(a), t' \in [\![\tau]\!]\}$, in which $[\![\tau]\!]$ denotes the interpretation of the τ by a set of hedges defined as follows:

$$[\![\emptyset]\!] = \emptyset \tag{4.16}$$

$$[\![\epsilon]\!] = \{\varepsilon\} \tag{4.17}$$

$$[\![a]\!] = \{o\langle\varepsilon\rangle \mid o \in \mathcal{O} \mid \omega(o) = \alpha(a)\} \tag{4.18}$$

$$[\![\tau_1 \mid \tau_2]\!] = [\![\tau_1]\!] \cup [\![\tau_2]\!] \tag{4.19}$$

$$[\![\tau_1\tau_2]\!] = \{\kappa(t_1, t_2) \mid t_i \in [\![\tau_i]\!] \text{ for } i = 1, 2\} \tag{4.20}$$

$$[\![\tau^*]\!] = \{\kappa(t_1, \kappa(t_2, \dots, \kappa(t_{n-1}, t_n) \dots)) \mid n \in \mathbb{N}, t_i \in [\![\tau]\!]\} \tag{4.21}$$

Most database queries and updates require a pair of database schemata to restrain input and output databases, respectively. Therefore, a tree-based background should provide at least two tree type schemes that are associated with initial and final states, respectively.

Definition 4.4.9. A *tree-based background* is a pair $(K^{tree}, \tilde{\Gamma}^{tree})$ consisting of

- a *tree background class* K^{tree}, and
- a set $\tilde{\Gamma}^{tree}$ of *tree type schemes.*

4.4.4 DB-ASMs for XML Database Transformations

Let us finally link the discussion of tree-based backgrounds with the postulates for database transformations and DB-ASMs as presented in Chapter 3. Following common practice, we treat XML database schemata and instances separately.

Definition 4.4.10. An *XML database schema* is a finite, non-empty set $\mathfrak{D}$ of EDTDs, while an *XML database instance* $I(\mathfrak{D})$ over $\mathfrak{D}$ is a finite, non-empty set of XML trees such that the following two conditions must both be satisfied:

- each XML tree $t_\gamma \in I(\mathfrak{D})$ is associated with an EDTD $G \in \mathfrak{D}$ such that $t_\gamma \in sat(G)$;
- each EDTD $G \in \mathfrak{D}$ has at least one XML tree $t_\gamma \in I(\mathfrak{D})$ such that $t_\gamma \in sat(G)$.

Thus, informally speaking, an XML database is a finite, unordered collection of XML trees, each of which should be associated with a tree name uniquely identifiable in a state, corresponding to the name of an XML document. Thus, we assume that in a state signature there is a finite, non-empty set of tree names, which are unary (and dynamic) function symbols.

Definition 4.4.11. An *XML database transformation* is a database transformation with a tree-based background $(K^{tree}, \tilde{\Gamma}^{tree})$ such that the following conditions are satisfied:

- the database part of the state signature contains a finite, non-empty set of unary function symbols representing tree names,
- the database part of each initial and final state satisfies the conditions of Definition 4.4.10 with EDTDs defined by the tree type schemes in $\tilde{\Gamma}^{tree}$.

In essence, an XML database transformation describes a process starting from some XML trees that are constrained by an input schema, processing them in accordance with database postulates defined in Section 3.1 and terminating with some XML trees that are constrained by an output schema.

The following theorem is a direct consequence of Theorems 3.2.1 and 3.3.1.

Theorem 4.4.2. *DB-ASMs with tree-based backgrounds* $(K^{tree}, \tilde{\Gamma}^{tree})$*, in which initial and final states satisfy the conditions of Definition 4.4.10 with EDTDs defined by the tree type schemes in* $\tilde{\Gamma}^{tree}$ *capture exactly all XML database transformations.*

Remark 4.4.1. For XML database transformations that do not impose any schema constraints on input and output XML database instances, Theorem 4.4.2 still holds by simply removing the conditions of Definition 4.4.10 with EDTDs defined by the tree type schemes in $\tilde{\Gamma}^{tree}$ for initial and final states. The reason of taking the schema information into consideration for XML database transformations is to show how such constraints can be captured in a tree-based background class. By working on this more restricted case, we can indeed easily extend our results to more general cases.

4.5 XML Machines

In this section, we present an alternative computational model for XML, which is called *XML machines*. The tree-based DB-ASM model from the previous section does not exploit weak MSO logic discussed in Section 4.3. The fact that it nonetheless captures all XML database transformation is mainly attributed to the power of DB-ASM rules. Using MSO_X, however, permits more sophisticated navigation over XML trees. Thus, one of the extensions of XML machines is the incorporation of MSO_X formulae in forall and choice rules. Besides this, we add partial update rules, which is purely for convenience as it does not add any additional expressive power. The other rules used by XML machines are the same as those in the tree-based DB-ASM model.

4.5.1 Extended Rules

As XML trees are unranked, a node may have an unbounded number of children nodes. To access all of them, we have two choices. One possibility is to process the children nodes sequentially one by one using an unbounded loop. The alternative is to execute in parallel an unbounded number of processes, one for each child. The former approach requires only standard total updates, whereas the latter one involves partial updates. For the sake of simplicity and naturalness of the computation model, the latter becomes our choice.

Using an unbounded number of parallel processes, we need an operator to merge two hedges into one. Using hedge juxtaposition by means of the algebric operation κ is one possibility, but we run into the problem that the order may not be the desired one. Therefore, we consider hedges as forests by ignoring the order and using simply forest union $\cup$. The non-determinism provided by choice rules can be exploited for this, i.e. choose any order for the resulting forest to turn it back into a hedge. In doing so, $\cup$ is defined over sort $\mathbf{H}$ (i.e. $\cup : \mathbf{H}^2 \to \mathbf{H}$) and becomes part of the background.

With these preliminary remarks, we can now define *MSO*-rules analogous to DB-ASM rules from Definition 3.2.1. In the following definition, the formulae φ always refer to formulae with trees as discussed in Section 4.3. That is, the logic MSO_X is included and equations between terms defined by the state signature can also be used.

Definition 4.5.1. The set $\mathcal{R}_{MSO}$ of *MSO-rules* over a signature $\Upsilon = \Upsilon_{db} \cup \Upsilon_a \cup \{f_1, \dots, f_\ell\}$ and a tree-based background K^{tree} is defined as follows:

- If t is a term over Υ and t' is a location in Υ such that $\#t' = \#t$, then

$$t' := t$$

  is a rule r in $\mathcal{R}_{MSO}$ called *assignment rule* with $fr(r) = var(t) \cup var(t')$, where $var(t)$ and $var(t')$ are the sets of variables occurring in the term t and the location t', respectively.

- If t is a term over Υ, t' is a location in Υ and $\cup$ is a binary operator such that $\#t' = \#t$ and $\cup : \#t \times \#t \to \#t'$, then

$$t' \leftleftharpoons^{\cup} t$$

is a rule r in $\mathcal{R}_{MSO}$ called *partial assignment rule* with $fr(r) = var(t) \cup var(t^{'})$, where $var(t)$ and $var(t^{'})$ are the sets of variables occurring in the term t and the location $t^{'}$, respectively.

- If φ is a formula and $r' \in \mathcal{R}_{MSO}$ is an MSO-rule, then

$$\textbf{if } \varphi \textbf{ then } r' \textbf{ endif}$$

is a rule r in $\mathcal{R}_{MSO}$ called *conditional rule* with $fr(r) = fr(\varphi) \cup fr(r')$.

- If φ is a formula with only database variables, $\{x_1, \dots, x_k, X_1, \dots, X_m\} \subseteq fr(\varphi)$ and $r' \in \mathcal{R}$ is an MSO-rule, then

$$\textbf{forall } x_1, \dots, x_k, X_1, \dots, X_m \textbf{ with } \varphi \textbf{ do } r' \textbf{ enddo}$$

is a rule r in $\mathcal{R}_{MSO}$ called *forall rule* with $fr(r) = (fr(r') \cup fr(\varphi)) - \{x_1, \dots, x_k, X_1, \dots, X_m\}$.

- If r_1, r_2 are rules in $\mathcal{R}_{MSO}$, then

$$\textbf{par } r_1 \dots r_n \textbf{ endpar}$$

is a rule r in $\mathcal{R}_{MSO}$, called *parallel rule* with $fr(r) = \bigcup_{1 \leq i \leq n} fr(r_i)$.

- If φ is a formula with only database variables, $\{x_1, \dots, x_k, X_1, \dots, X_m\} \subseteq fr(\varphi)$ and $r' \in \mathcal{R}_{MSO}$ is an MSO-rule, then

$$\textbf{choose } x_1, \dots, x_k, X_1, \dots, X_m \textbf{ with } \varphi \textbf{ do } r' \textbf{ enddo}$$

is an MSO-rule r in $\mathcal{R}_{MSO}$ called *choice rule* with $fr(r) = (fr(r') \cup fr(\varphi)) - \{x_1, \dots, x_k, X_1, \dots, X_m\}$.

- If r_1, r_2 are rules in $\mathcal{R}_{MSO}$, then

$$\textbf{seq } r_1 \; r_2 \textbf{ endseq}$$

is a rule r in $\mathcal{R}_{MSO}$, called *sequence rule* with $fr(r) = fr(r_1) \cup fr(r_2)$.

- If r' is a rule in $\mathcal{R}_{MSO}$ and θ is a location function that assigns location operators ρ to terms t with $var(t) \subseteq fr(r')$, then

$$\textbf{let } \theta(t) = \rho \textbf{ in } r' \textbf{ endlet}$$

is a rule r in $\mathcal{R}_{MSO}$ called *let rule* with $fr(r) = fr(r')$.

The definition of associated sets of update sets $\Delta(r, S)$ for a closed MSO-rule r with respect to a state S is again straightforward. We only explain the non-standard case of partial assignment rules.

For this, let S be a state over Υ and ζ be a variable assignment for $fr(r)$. We then obtain

$$\Delta(t' \leftleftharpoons^{\cup} t, S, \zeta) = \{\{(\ell, a, \cup)\}\}$$

with $\ell = val_{S,\zeta}(t')$ and $a = val_{S,\zeta}(t)$. That is, we obtain a single update set with a single partial assignment to the location ℓ. As this rule r can appear as part of a complex MSO-rule without free variables, the variable assignment ζ will be determined by the context and the partial update will become an element of larger update sets Δ. Then, for a state S, the value of location ℓ in the successor state $S + \Delta$ becomes

$$val_{S+\Delta}(\ell) \;=\; val_S(\ell) \cup \bigcup_{(\ell,v,\cup)\in\Delta} v\,,$$

if the value on the right hand side is defined unambiguously, otherwise $val_{S+\Delta}(\ell)$ will be undefined.

Example 4.5.1. Consider the XML tree (i) in Figure 4.4 and assume it is assigned to the variable (tree name) x_{exa}. The following MSO-rule will construct the XML tree in (ii) from subtrees of the given XML tree:

$t_1 := \epsilon$;

forall x,y,z **with** $\prec_c (x_{exa}, root(x_{exa}), x) \wedge \prec_c (x_{exa}, x, y) \wedge \prec_c (x_{exa}, x, z)$
$\wedge \omega(x_{exa}, x) = b \wedge \omega(x_{exa}, y) = c \wedge \omega(x_{exa}, z) = a$

do

$t_1 \leftleftharpoons^{\cup} f_\iota(c, f_\kappa(subtree(x_{exa}, y), subtree(x_{exa}, z)))$;

enddo ;

$output := f_\iota(r, t_1)$

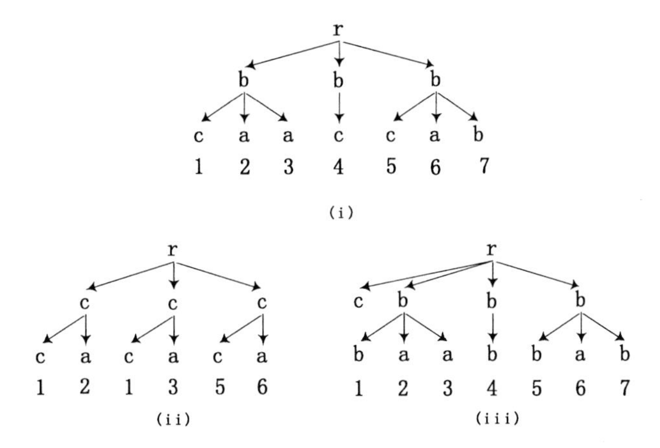

Figure 4.4: An XML tree and results of tree operations

Similarly, the XML tree (iii) in Figure 4.4 is obtained after executing the following MSO-rule:

$t_1 := \epsilon$;
forall x,y **with** $\prec_c (x_{exa}, root(x_{exa}), x) \wedge \prec_c (x_{exa}, x, y) \wedge$
$\omega(x_{exa}, x) = b \wedge \omega(x_{exa}, y) = c$
do
$t_1 \leftleftarrows^{\cup} f_\eta(context(x_{exa}, x, y), f_\iota(b, \varepsilon))$;
enddo ;
$output := f_\iota(r, f_\kappa(f_\iota(c, \varepsilon), t_1))$

□

Example 4.5.2. The following MSO-rule constructs the XML tree in Figure 4.5 from subtrees of the given XML tree in Figure 4.4(i), each of which is rooted at a node labeled as b with at least two descendant nodes labeled as a and c, respectively.

$t_1 := \epsilon$;
forall x **with** $\omega(t_{exa}, x) = b \wedge \exists X.(\forall x_1, x_2.((x_1 \in X \wedge \prec_c (t_{exa}, x_1, x_2)$
$\Rightarrow x_2 \in X) \wedge \forall x_1.(\prec_c (t_{exa}, x, x_1) \Rightarrow x_1 \in X)))$

$\wedge \exists y, z.(\omega(t_{exa}, y) = c \wedge \omega(t_{exa}, z) = a \wedge y \in X \wedge z \in X)$

do $t_1 \leftleftarrows^{\cup} subtree(t_{exa}, x)$ **enddo**;

$output := \iota(d, t_1)$

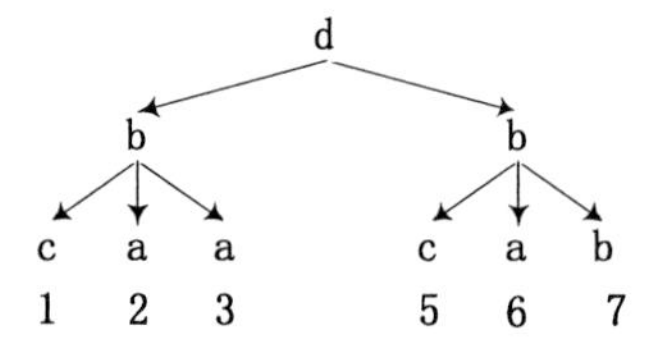

Figure 4.5: Another result of tree operations

□

Definition 4.5.2. An *XML Machine* M over signature Υ as in Postulate 3.1.2 with a tree-based background $(K^{tree}, \tilde{\Gamma}^{tree})$ as in Definition 4.4.9 consists of

- a set $\mathcal{S}_M$ of states over Υ satisfying the requirements in Postulate 3.1.2 and closed under isomorphisms,
- non-empty subsets $\mathcal{I}_M \subseteq \mathcal{S}_M$ of initial states, and $\mathcal{F}_M \subseteq \mathcal{S}_M$ of final states, both also closed under isomorphisms and satisfying the conditions of Definition 4.4.10 with EDTDs defined by the tree type schemes in $\tilde{\Gamma}^{tree}$,
- a program p_M defined by a closed MSO-rule r over Υ, and
- a binary relation δ_M over $\mathcal{S}_M$ determined by p_M such that

$$\{S_{i+1} \mid (S_i, S_{i+1}) \in \delta_M\} = \{S_i + \Delta \mid \Delta \in \Delta(p_M, S_i)\}$$

holds.

4.5.2 Behavioural Equivalence

In the following we will show the behavioural equivalence between DB-ASMs with tree-based backgrounds and the computation model for XML (i.e., XML Machines) as introduced in the last subsection.

Theorem 4.5.1. *The XML Machines with a tree-based background* $(K^{tree}, \tilde{\Gamma}^{tree})$ *capture exactly all XML database transformations with the same signature and background.*

Proof. According to Theorem 4.4.2, each XML database transformation can be represented by a behaviourally equivalent DB-ASM with the same tree-based background and signature, and vice versa. Since DB-ASMs differ from XML Machines only by the fact that DB-ASM rules are more restrictive than MSO-rules (they do not permit MSO_X formulae in forall and choice rules), such a DB-ASM is in fact also an XML Machine.

Thus, it suffices to show that XML Machines satisfy the postulates for XML database transformations defined in Postulates 3.1.1 (sequential time postulate), 3.1.2 (abstract state postulate), 3.1.3 (background postulate), 3.1.4 (bounded exploration postulate) and 3.1.5 (bounded non-determinism postulate). The first three of these postulates are already captured by the definitions of XML Machines and tree-based background, so we have to consider only the bounded exploration and bounded non-determinism postulates.

Regarding bounded exploration we note that the assignment rules within the MSO-rule r that defines p_M are decisive for the set of update set $\Delta(r, S)$ for any state S. Hence, if $f(t_1, \dots, t_n) := t_0$ is an assignment rule occurring within r, and $val_{S,\zeta}(t_i) = val_{S',\zeta}(t_i)$ holds for all $i = 0, \dots, n$ and all variable assignments ζ that have to be considered, then we obtain $\Delta(r, S) = \Delta(r, S')$.

We use this to define a bounded exploration witness $\mathcal{T}_{wit}$. If t_i is ground, we add the access term t_i to $\mathcal{T}_{wit}$. If t_i is not ground, then the corresponding assignment rule must appear within the scope of the forall and choice rules introducing the database variables in t_i, as r is closed. Thus, variables in t_i are bound by a formula φ. That is, for $fr(t_i) = \{x_1, \dots, x_k\}$ the relevant variable assignments are $\zeta = \{x_1 \mapsto b_1, \dots, x_k \mapsto b_k\}$ with $val_{S,\zeta}(\varphi) = true$. Bringing φ into a form that only uses conjunction, negation and existential quantification, we can extract a set of access terms $\{(t_{\beta_1}, t_{\alpha_1}), \dots, (t_{\beta_\ell}, t_{\alpha_\ell})\}$ such that if S and S' coincide on these access terms, they will also coincide on the formula φ. This is possible, as we evaluate access terms by sets, so conjunction corresponds to union, existential quantification to projection, and negation to building the (finite) complement. We add all the access terms $(t_{\beta_1}, t_{\alpha_1}), \dots, (t_{\beta_\ell}, t_{\alpha_\ell})$ to $\mathcal{T}_{wit}$.

More precisely, if φ is a conjunction $\varphi_1 \wedge \varphi_2$, then $\Delta(r, S_1) = \Delta(r, S_2)$ will hold, if $\{(b_1, \dots, b_k) \mid val_{S_1,\zeta}(\varphi) = true\} = \{(b_1, \dots, b_k) \mid val_{S_2,\zeta}(\varphi) = true\}$ holds (with $\zeta = \{x_1 \mapsto b_1, \dots, x_k \mapsto b_k\}$). If T_i is a set of access terms such that whenever S_1 and S_2 coincide on T_i, then $\{(b_1, \dots, b_k) \mid val_{S_1,\zeta}(\varphi_i) = true\} = \{(b_1, \dots, b_k) \mid val_{S_2,\zeta}(\varphi_i) = true\}$ will hold $(i = 1, 2)$, then $T_1 \cup T_2$ is a set of access terms such that whenever S_1 and S_2 coincide on $T_1 \cup T_2$, then $\{(b_1, \dots, b_k) \mid val_{S_1,\varsigma}(\varphi) = true\} = \{(b_1, \dots, b_k) \mid val_{S_2,\varsigma}(\varphi) = true\}$ will hold.

Similarly, a set of access terms for ψ with the desired property will also be a witness for $\varphi = \neg\psi$, and $\bigcup_{b_{k+1} \in B_{db}} T_{b_{k+1}}$ with sets of access terms $T_{b_{k+1}}$ for $\psi[t_{k+1}/x_{k+1}]$ with $val_S(t_{k+1}) = b_{k+1}$ defines a finite set of access terms for $\varphi = \exists x_{k+1}\psi$. In this way, we can restrict ourselves to atomic formulae, which are equations and thus give rise to canonical access terms.

Then by construction, if S and S' coincide on $\mathcal{T}_{wit}$, we obtain $\Delta(r, S) = \Delta(r, S')$. As there are only a finite number of assignments rules within r and only a finite number of choice and forall rules defining the variables in such assignments, the set $\mathcal{T}_{wit}$ of access terms must be finite, i.e. r satisfies the bounded exploration postulate.

Regarding bounded non-determinism, assuming that XML machines do not satisfy the bounded non-determinism postulate. It means that there does not exist a non-ground access term (t_β, t_α) in $\mathcal{T}_{wit}$. According to our remark above r must contain a choice rule **choose** $x_1, \dots, x_k$ **with** φ **do** r' **enddo**. Hence, it implies that there exist at least one non-ground access term in $\mathcal{T}_{wit}$ contradicting our assumption. □

Chapter 5

Database Transformation Logic

In this chapter our task is to investigate the logical characterisation of DB-ASMs, and thereby to define a logic for DB-ASMs. Particularly, we establish a sound and complete proof system for such a logic. Recall that, in Chapter 3, it has been shown that a database transformation is behaviourally equivalent to a DB-ASM and vice versa. Therefore, a rigorous proof system for the logic for DB-ASMs can offer vast advantages in reasoning about database transformations, such as verifying the correctness of specification, deriving static or dynamic properties, determining the equivalence of programs, comparing the expressive power of computation models, etc.

Since states of a database transformation are meta-finite structures as stipulated by the abstract state postulate (i.e., Postulate 3.1.2), we begin by introducing the logic of meta-finite states in Section 5.1. In its most general form, the logic of meta-finite states is parameterised by a logic for finite structures, and its syntax and semantics are presented in Subsections 5.1.1 and 5.1.2, respectively. Due to the presence of ρ-terms which formalise multiset-operations in the algorithmic part of a meta-finite state, the expressive power of the logic for meta-finite states has been greatly enhanced.

In Section 5.2 we define a logic for DB-ASMs. The syntax and semantics are provided in Subsections 5.2.1 and 5.2.1, which correspondingly extend the syntax and semantics of the logic of meta-finite states. Subsequently, a detailed discussion on various properties of the logic for DB-ASMs, such as, non-determinism, consistency, definedness, and update sets and multisets, is presented in Subsections 5.2.3 - 5.2.7. The most challenging problem we face is the handling of non-deterministic update sets associated with DB-ASM rules. This is resolved by the addition of the modal operator [ ] for an update set. The approach works well because update sets yielded by DB-ASM rules are assured to be finite.

The formalisation of a proof system for the logic for DB-ASMs is presented in Section 5.3. In spite of the inclusion of second-order formulae in the logic for DB-ASMs, the finiteness of domains that quantifiers are restricted to allows us to show that there exists a transition from the logic for DB-ASMs to the first-order logic in Subsection 5.5.1. Similar to the hierarchical ASMs which have no cycles in the dependency graph of rule declarations, there is no recursion in any DB-ASM rule. Therefore, we are able to present two alternative completeness proofs for the logic for DB-ASMs: by Henkin construction and by transforming into definitional extension of the first-order logic. These two approaches are elaborated in Subsections 5.5.2 and 5.5.3, respectively.

5.1 A Logic of Meta-finite States

In [74] Grädel and Gurevich extended a logic suitable for finite structures (e.g., first-order logic, fixed point logic, the infinitary logic, etc.) to a logic of meta-finite structures. They aimed at investigating the logical characterisation of meta-finite structures. On top of their work, Hella et al. [93] extended an infinitary counting logic from Libkin [106] for the purpose of studying the logical grounds of query languages with aggregation. In this section we adopt the logic of meta-finite structures from [74] to characterise states of a database transformation.

5.1.1 Syntax

Let us fix a logic $\mathfrak{L}$ of finite structures and a countable set $\mathcal{X}_{FO} = \{x_1, x_2, ...\}$ of first-order variables. For simplicity, $\overline{x}$, $\overline{y}$,... are used to denote tuples of variables in $\mathcal{X}_{FO}$.

Definition 5.1.1. Let $\Upsilon = \Upsilon_{db} \cup \Upsilon_a \cup \{f_1, \dots f_\ell\}$ be a signature of meta-finite states, $\{\rho_1, ..., \rho_m\} \subseteq \Upsilon_a$ be a set of location operators, and the variables in $\mathcal{X}_{FO}$ be restricted to range only over the database part of a meta-finite state (i.e., a finite structure as defined in Postulate 3.1.2). The terms and formulae of the logic $\mathbb{L}(\Upsilon, \mathfrak{L})$ over meta-finite states with signature Υ which is parameterised by the logic $\mathfrak{L}$ are inductively defined by the following rules.

- The set $\mathcal{T}$ of terms is constituted by the sets $\mathcal{T}_{db}$ and $\mathcal{T}_a$ of terms over the database and algorithmic parts called *database terms* and *algorithmic terms*, respectively (i.e., $\mathcal{T} = \mathcal{T}_{db} \cup \mathcal{T}_a$), such that

(1) $x \in \mathcal{T}_{db}$, where $x \in \mathcal{X}_{FO}$ and $fr(x) = \{x\}$;

(2) $f(t_1, ..., t_n) \in \mathcal{T}_{db}$, where $f \in \Upsilon_{db}$ is a n-ary function symbol, $\{t_1, ..., t_n\} \subseteq \mathcal{T}_{db}$, and $fr(f(t_1, ..., t_n)) = \bigcup_{1 \leq j \leq n} fr(t_j)$;

(3) $f(t_1, ..., t_n) \in \mathcal{T}_a$, where $f \in \{f_1, \dots f_\ell\}$ is a n-ary function symbol, $\{t_1, ..., t_n\} \subseteq \mathcal{T}_{db}$ and $fr(f(t_1, ..., t_n)) = \bigcup_{1 \leq j \leq n} fr(t_j)$;

(4) $f(t_1, ..., t_n) \in \mathcal{T}_a$, where $f \in \Upsilon_a - \{\rho_1, ..., \rho_m\}$ is a n-ary function symbol, $\{t_1, ..., t_n\} \subseteq \mathcal{T}_a$, and $fr(f(t_1, ..., t_n)) = \bigcup_{1 \leq j \leq n} fr(t_j)$;

(5) $\rho_{\overline{x}}(t|\varphi(\overline{x}, \overline{y})) \in \mathcal{T}_a$, where $t \in \mathcal{T}_a$, $\rho \in \{\rho_1, ..., \rho_m\}$, $\varphi(\overline{x}, \overline{y}) \in \Phi$ is a formula defined in the following with $fr(t) \subseteq fr(\varphi(\overline{x}, \overline{y}))$ and $fr(\rho_{\overline{x}}(t|\varphi(\overline{x}, \overline{y}))) = \overline{y}$.

- The set Φ of formulae consists of

  (i) atomic formulae:

  - $t_1 = t_2$ for either $t_1, t_2 \in \mathcal{T}_{db}$ or $t_1, t_2 \in \mathcal{T}_a$, and $fr(t_1 = t_2) = fr(t_1) \cup fr(t_2)$;
  - $P_1(t_1, ..., t_n)$ for n-ary predicate symbol $P_1 \in \Upsilon_{db}$, $t_i \in \mathcal{T}_{db}$ $(i = 1, ..., n)$, and $fr(P_1(t_1, ..., t_n)) = \bigcup_{1 \leq j \leq n} fr(t_j)$;
  - $P_2(t_1, ..., t_n)$ for n-ary predicate symbol $P_2 \in \Upsilon_a$, $t_i \in \mathcal{T}_a$ $(i = 1, ..., n)$, and $fr(P_2(t_1, ..., t_n)) = \bigcup_{1 \leq j \leq n} fr(t_j)$;

  (ii) the set of formulae closed under all rules of the logic $\mathfrak{L}$ for building formulae, with the restriction on all variables in these formulae such that variables are only permitted to range over the database part.

For the convenience of expression, the notations *ρ-term* referring to a term in the form of $\rho_{\overline{x}}(t|\varphi(\overline{x}, \overline{y}))$ and *pure term* referring to a term defined by only applying Rules (1)-(4) (i.e., terms that do not contain any formulae and ρ-terms) are used. As defined in Rule (5), ρ-terms are constructed on top of formulae, and can also be part of other formulae. This may lead to an iterative creation between formulae and terms. In order to formally express the nesting depth of ρ-terms, we associate a rank, called *ρ-rank*, with each term in $\mathbb{L}(\Upsilon, \mathfrak{L})$ to describe the nesting depth of ρ-terms such that

- pure terms have ρ-rank 0;

- ρ-terms $\rho_{\overline{x}}(t|\varphi(\overline{x},\overline{y}))$ have ρ-rank $n+1$ if the maximal ρ-rank of t and terms in $\varphi(\overline{x},\overline{y})$ is n;
- terms built upon ρ-terms have ρ-rank n if the maximal ρ-rank of these ρ-terms is n.

5.1.2 Semantics

Assume that the notation $[\overline{x} \mapsto \bar{a}]$ is used as a shorthand for $[x_1 \mapsto a_1, \ldots, x_n \mapsto a_n]$, where $\overline{x} = (x_1, \ldots, x_n)$ is a tuple of variables and $\bar{a} = (a_1, \ldots, a_n)$ is a tuple of constants.

Definition 5.1.2. Let S be a meta-finite state of signature Υ with the base set $B = B_{db} \cup B_a$, and ζ be a variable assignment. The semantics for the terms and formulae of the logic $\mathbb{L}(\Upsilon, \mathfrak{L})$ is inductively defined by

terms :

- $val_{S,\zeta[x\mapsto a]}(x) = a$ for $a \in B_{db}$
- $val_{S,\zeta}(f(t_1,\ldots,t_n)) = f(val_{S,\zeta}(t_1),\ldots,val_{S,\zeta}(t_n))$ for $j = 1,\ldots,n$;
  - if $f \in \Upsilon_{db}$, then $val_{S,\zeta}(t_i)$ and $f(val_{S,\zeta}(t_1),\ldots,val_{S,\zeta}(t_n)) \in B_{db}$ for $i = 1,\ldots,n$;
  - if $f \in \Upsilon_a$, then $val_{S,\zeta}(t_i)$ and $f(val_{S,\zeta}(t_1),\ldots,val_{S,\zeta}(t_n)) \in B_a$ for $i = 1,\ldots,n$;
  - if $f \in \{f_1,\ldots,f_\ell\}$, then $val_{S,\zeta}(t_i) \in B_{db}$ for $i = 1,\ldots,n$ and $f(val_{S,\zeta}(t_1),\ldots,val_{S,\zeta}(t_n)) \in B_a$
- $val_{S,\zeta[\overline{y}\mapsto\bar{b}]}(\rho_{\overline{x}}(t|\varphi(\overline{x},\overline{y}))) =$
$$\begin{cases} \rho(\{\!\{val_{S,\zeta[\overline{y}\mapsto\bar{b},\overline{x}\mapsto\bar{a}]}(t) | \text{ for all } \bar{a} \text{ such that } [\![\varphi(\overline{x},\overline{y})]\!]_{S,\zeta[\overline{y}\mapsto\bar{b},\overline{x}\mapsto\bar{a}]} = true\}\!\}) \\ \quad \text{if there exists at least one } \bar{a} \text{ such that } [\![\varphi(\overline{x},\overline{y})]\!]_{S,\zeta[\overline{y}\mapsto\bar{b},\overline{x}\mapsto\bar{a}]} = true; \\ \bot \text{ otherwise.} \end{cases}$$

formulae :

- $[\![t_1 = t_2]\!]_{S,\zeta} = \begin{cases} true & \text{if } val_{S,\zeta}(t_1) = val_{S,\zeta}(t_2) \\ false & \text{otherwise} \end{cases}$
- $[\![P_i(t_1,...,t_n)]\!]_{S,\zeta} = \begin{cases} true & \text{if } (val_{S,\zeta}(t_1),...,val_{S,\zeta}(t_n)) \in P_i \\ false & \text{otherwise} \end{cases}$ for $i = 1, 2$

Due to the restriction that all variables in $\mathbb{L}(\Upsilon, \mathfrak{L})$ can only range over B_{db}, the finiteness condition on the database part of a meta-finite state defined in the abstract state postulate (i.e., Postulate 3.1.2) then implies that every multiset $\{\!\{val_{S,\zeta[\bar{y}\mapsto\bar{b},\bar{x}\mapsto\bar{a}]}(t)|$ for all $\bar{a}$ such that $[\![\varphi(\bar{x},\bar{y})]\!]_{S,\zeta[\bar{y}\mapsto\bar{b},\bar{x}\mapsto\bar{a}]} = true\}\!\}$ has a finite number of elements. Nevertheless, as ρ-terms are built upon formulae by recursively applying Rules (1)-(5) and (i)-(ii) in Definition 5.1.1, the logic $\mathbb{L}(\Upsilon, \mathfrak{L})$ is indeed very powerful. The following example illustrates the expressive power of $\mathbb{L}(\Upsilon, \mathfrak{L})$ in aggregate computing of database applications.

Example 5.1.1. Let us consider a meta-finite state as described in Example 3.1.1. Recall that there exists a relation schema AUTHORSHIP = {PubID,UnitID,PersonID,Order} in the database part. We also assume that the location operators *max* and *sum* are contained in the algorithmic part. Then, the following two aggregate queries are able to be expressed in $\mathbb{L}(\Upsilon, \mathfrak{L})$.

Q_1: Calculate the total number of publications in the database.

$$sum_{x_1}(1|\exists x_2, x_3, x_4.\text{AUTHORSHIP}(x_1, x_2, x_3, x_4))$$

Alternatively, Q_1 can be expressed by the following SQL statement:

SELECT *sum*(cnum) FROM

(SELECT *count*(distinct PubID) AS cnum, PubID FROM

AUTHORSHIP

GROUP BY PubID)

Q_2: Find the author who has published the maximal number of publications in the database.

$$max_{x_3}(sum_{x_1}(1|\exists x_2, x_4.\text{AUTHORSHIP}(x_1, x_2, x_3, x_4))$$
$$|\exists y_1, y_2, y_4.\text{AUTHORSHIP}(y_1, y_2, x_3, y_4))$$

In a similar way, Q_2 can be expressed by the following SQL statement:

SELECT *max*(snum) FROM

(SELECT snum, a.PersonID FROM

AUTHORSHIP a,

(SELECT *sum*(cnum) AS snum, PersonID FROM

(SELECT *count*(distinct PubID) AS cnum, PersonID, PubID FROM

AUTHORSHIP

GROUP BY PubID, PersonID)

GROUP BY PersonID) b

WHERE a.PersonID=b.PersonID

GROUP BY a.PersonID)

□

Grädel and Gurevich [74] provided some examples showing that several computations in an arithmetical structure such as counting equivalence classes, binary representations of natural numbers, etc. are also definable in a logic of meta-finite states.

5.2 A Logic for DB-ASMs

We provide a logical characterisation for DB-ASMs in this section. The logic for DB-ASMs (denoted as $\mathbb{L}^{\mathfrak{L}}_{M}$) is a logic built upon the logic $\mathbb{L}(\Upsilon, \mathfrak{L})$ of meta-finite states, in which DB-ASMs and meta-finite states have the same signature Υ. Our approach to develop the logic $\mathbb{L}^{\mathfrak{L}}_{M}$ is in the same spirit of the logic for ASMs as defined in [47], except for the following distinctions.

- First of all, DB-ASMs are able to collect updates yielded in parallel computations under the multiset semantics, i.e., update multisets, and then aggregate updates in an update multiset to an update set by using location operators. These distinguished features of DB-ASMs are captured by the logic for DB-ASMs via the use of ρ-terms.

- Secondly, due to the importance of non-determinism for enhancing the expressive power of database transformations, DB-ASMs take into account choice rules. Thus, as a logic for DB-ASMs, $\mathbb{L}^{\mathfrak{L}}_{M}$ has to handle all the issues around non-determinism that have been identified as the source of problems in the completeness proof of the logic for ASMs [47].

- The last difference is more of syntactical nature. To obtain more concise and natural expressions in the logic $\mathbb{L}_M^{\mathfrak{L}}$, we make update sets and multisets explicit in the formalisation of $\mathbb{L}_M^{\mathfrak{L}}$; furthermore, second-order variables are used by being bounded to update sets or multisets. Nevertheless, restricted by the finiteness condition on updates in an update set or multiset, when the logic $\mathbb{L}(\Upsilon, \mathfrak{L})$ of meta-finite states is parameterised by the first-order logic, i.e., $\mathfrak{L}$=FO, $\mathbb{L}_M^{FO}$ is not more expressive than the first-order logic.

5.2.1 Syntax

Extending the syntax of the logic $\mathbb{L}(\Upsilon, \mathfrak{L})$ of meta-finite states we introduced in the previous section, the syntax of the logic $\mathbb{L}_M^{\mathfrak{L}}$ for DB-ASMs can be formalised as follows.

Definition 5.2.1. Let M be a DB-ASM over signature Υ. The set $\mathcal{T}_M$ of terms in the logic $\mathbb{L}_M^{\mathfrak{L}}$ for DB-ASMs is defined in the same way as the set of terms in $\mathbb{L}(\Upsilon, \mathfrak{L})$ as defined in Definition 5.5.1, and the set Φ_M of formulae in the logic $\mathbb{L}_M^{\mathfrak{L}}$ for DB-ASMs comprises all formulae in $\mathbb{L}(\Upsilon, \mathfrak{L})$ as defined in Definition 5.5.1, and also the following extended formulae:

$$\begin{aligned}\varphi :\equiv\ & \exists X.\varphi \,|\, \forall X.\varphi \,| \\ & \text{upd}(r, \Delta) \,|\, \text{upm}(r, \ddot{\Delta}) \,| \\ & \Delta(f, \bar{t}, t_0) \,|\, \ddot{\Delta}(f, \bar{t}, t_0, t') \,| \\ & \text{def}(r) \,|\, [\Delta]\varphi\end{aligned}$$

The formulae $\exists X.\varphi$ and $\forall X.\varphi$ in $\mathbb{L}_M^{\mathfrak{L}}$ are second-order formulae in which X is a second-order variable bound to an update set Δ or an update multiset $\ddot{\Delta}$. The predicates upd(r, Δ) and upm$(r, \ddot{\Delta})$ describe an update set Δ and an update multiset $\ddot{\Delta}$ generated by a rule r of M, respectively. As discussed in Chapter 3, when applying forall and parallel rules of a DB-ASM, updates produced in parallel computations may be identical and thus need the multiset semantics. For this reason, both predicates upd(r, Δ) and upm$(r, \ddot{\Delta})$ are included in $\mathbb{L}_M^{\mathfrak{L}}$. As for the predicates $\Delta(f, \bar{t}, t_0)$ and $\ddot{\Delta}(f, \bar{t}, t_0, t')$, $\Delta(f, \bar{t}, t_0)$ describes that an update $(f(\bar{t}), t_0)$ exists in an update set Δ, while $\ddot{\Delta}(f, \bar{t}, t_0, t')$ describes that an update $(f(\bar{t}), t_0)$ is the t'th occurrence in an update multiset $\ddot{\Delta}$. The predicate def(r) is used to formulate the definedness property of a rule r of M the same as used in [47]. Instead of introducing modal operators $[\,]$ and $\langle\,\rangle$ for a rule r, i.e., the formulae $[r]\varphi$ and $\langle r\rangle\varphi$ expressing the evaluation of φ over a state after executing the rule r on the current state, we use $[\Delta]\varphi$ to express the evaluation of φ over all states after executing the update set Δ on the current state. The connections between $[r]\varphi$, $\langle r\rangle\varphi$ and $[\Delta]\varphi$ will be further explained in the next subsection.

A formula is *pure* if it does not contain any extended formulae, i.e., $\exists X.\varphi$, $\forall X.\varphi$, $\text{upd}(r,\Delta)$, $\text{upm}(r,\ddot{\Delta})$, $\Delta(f,\bar{t},t_0)$, $\ddot{\Delta}(f,\bar{t},t_0,t')$, $\text{def}(r)$ and $[\Delta]\varphi$. A formula or a term is *static* if it does not contain any dynamic function symbols. The formulae occurring in conditional, forall and choice rules of a DB-ASM M must be pure formulae in $\mathbb{L}^{\mathfrak{L}}_{M}$.

5.2.2 Semantics

We define the semantics for the extended formulae in the logic $\mathbb{L}^{\mathfrak{L}}_{M}$ for DB-ASMs.

Definition 5.2.2. Let S be a meta-finite state of signature Υ, $\mathbb{F}_{dyn}$ be the set of all dynamic function symbols in signature Υ and ζ be a variable assignment. Then the semantics for the extended formulae of $\mathbb{L}^{\mathfrak{L}}_{M}$ is defined by

- $$[\![\exists X.\varphi]\!]_{S,\zeta} = \begin{cases} true & \text{if } [\![\varphi]\!]_{S,\zeta[X\mapsto P]} = true \text{ for some finite } P \subseteq \mathbb{F}_{dyn} \times B^n \times B \\ & \text{or } P \subseteq \mathbb{F}_{dyn} \times B^n \times B \times \mathbb{N} \text{ (n is determined by X),} \\ false & \text{otherwise} \end{cases}$$

- $$[\![\forall X.\varphi]\!]_{S,\zeta} = \begin{cases} true & \text{if } [\![\varphi]\!]_{S,\zeta[X\mapsto P]} = true \text{ for all finite } P \subseteq \mathbb{F}_{dyn} \times B^n \times B \\ & \text{or } P \subseteq \mathbb{F}_{dyn} \times B^n \times B \times \mathbb{N} \text{ (n is determined by X),} \\ false & \text{otherwise} \end{cases}$$

- $$[\![\text{upd}(r,\Delta)]\!]_{S,\zeta} = \begin{cases} true & \text{if } val_{S,\zeta}(\Delta) \in \Delta(r,S,\zeta), \\ false & \text{otherwise} \end{cases}$$

- $$[\![\text{upm}(r,\ddot{\Delta})]\!]_{S,\zeta} = \begin{cases} true & \text{if } val_{S,\zeta}(\ddot{\Delta}) \in \ddot{\Delta}(r,S,\zeta), \\ false & \text{otherwise} \end{cases}$$

- $$[\![\Delta(f,\bar{t},t_0)]\!]_{S,\zeta} = \begin{cases} true & \text{if } (f, val_{S,\zeta}(\bar{t}), val_{S,\zeta}(t_0)) \in val_{S,\zeta}(\Delta), \\ false & \text{otherwise} \end{cases}$$

- $$[\![\ddot{\Delta}(f,\bar{t},t_0,t')]\!]_{S,\zeta} = \begin{cases} true & \text{if } (f, val_{S,\zeta}(\bar{t}), val_{S,\zeta}(t_0), n) \in val_{S,\zeta}(\ddot{\Delta}) \text{ for some } n \in \mathbb{N} \\ & \text{and } val_{S,\zeta}(t') < n, \\ false & \text{otherwise} \end{cases}$$

- $[\![\text{def}(r)]\!]_{S,\zeta} = \begin{cases} true & \text{if } \Delta(r,S,\zeta) \neq \emptyset, \\ false & \text{otherwise} \end{cases}$

- $[\![[\Delta]\varphi]\!]_{S,\zeta} = \begin{cases} true & \text{if } [\![\varphi]\!]_{S+\Delta,\zeta} = true \text{ for each state after applying } \Delta \in \Delta(r,S,\zeta) \\ & \text{over state } S \\ false & \text{otherwise} \end{cases}$

In $\mathbb{L}^{\mathfrak{L}}_{M}$ we also use the expressions $\psi_1 \vee \psi_2$, $\psi_1 \Rightarrow \psi_2$, $\psi_1 \Leftrightarrow \psi_2$ and $\forall \overline{x}.\varphi$ as shortcuts in the standard way. For the modal expressions $[r]\varphi$ and $\langle r \rangle \varphi$ with the following semantics:

- $[\![[r]\varphi]\!]_{S,\zeta} = \begin{cases} true & \text{if } [\![\varphi]\!]_{S+\Delta,\zeta} = true \text{ for all consistent } \Delta \in \Delta(r,S,\zeta), \\ false & \text{otherwise} \end{cases}$

- $[\![\langle r \rangle \varphi]\!]_{S,\zeta} = \begin{cases} false & \text{if } [\![\varphi]\!]_{S+\Delta,\zeta} = false \text{ for all consistent } \Delta \in \Delta(r,S,\zeta), \\ true & \text{otherwise} \end{cases}$

they can be treated as the shortcuts in $\mathbb{L}^{\mathfrak{L}}_{M}$ with the following logical equivalences:

$$[r]\varphi \equiv \forall \Delta.(\text{upd}(r,\Delta) \Rightarrow [\Delta]\varphi),$$

$$\langle r \rangle \varphi \equiv \exists \Delta.(\text{upd}(r,\Delta) \wedge [\Delta]\varphi).$$

If Δ is inconsistent, then the formula $[\Delta]\varphi$ is interpreted as *true* in accordance with its semantics in Definition 5.2.2. The reason for this is that there is no successor states to the current state after applying an inconsistent update set over it. Therefore, the formula $[\Delta]\varphi$ in $\mathbb{L}^{\mathfrak{L}}_{M}$ can be treated in a very similar way to the formula $[r]\varphi$ in the logic for ASMs [134].

Remark 5.2.1. In $\mathbb{L}^{\mathfrak{L}}_{M}$ less-than-or-equal (denoted as $\leq$) and membership (denoted as $\in$) are used as predicates under a fixed interpretation defined in the standard way. Although the second-order formulae $\exists X.\varphi$ and $\forall X.\varphi$ are included in $\mathbb{L}^{\mathfrak{L}}_{M}$, a second-order variable X is always bound to an update set Δ or an update multiset $\ddot{\Delta}$ that must be finite.

More precisely, the finiteness of Δ (denoted as fin(Δ)) can be formulated as

- $\text{fin}(\Delta) \Leftrightarrow \bigwedge_{f \in \mathbb{F}_{dyn}} \exists \overline{x}_1, ..., \overline{x}_n, y_1, ..., y_n.(\Delta(f, \overline{x}_1, y_1) \wedge ... \wedge \Delta(f, \overline{x}_n, y_n) \wedge$
  $\forall \overline{x}, y.(\Delta(f, \overline{x}, y) \Rightarrow \bigvee_{1 \leq i \leq n} (\overline{x} = \overline{x}_i \wedge y = y_i))),$

and similarly, the finiteness of $\ddot{\Delta}$ (denoted as $\text{fin}(\ddot{\Delta})$) can be formulated as

- $\text{fin}(\ddot{\Delta}) \Leftrightarrow \bigwedge_{f \in \mathbb{F}_{dyn}} \exists \overline{x}_1, ..., \overline{x}_n, y_1, ..., y_n, z_1, ..., z_n.(\ddot{\Delta}(f, \overline{x}_1, y_1, z_1) \wedge ... \wedge \ddot{\Delta}(f, \overline{x}_n, y_n, z_n) \wedge$

  $\forall \overline{x}, y, z.(\ddot{\Delta}(f, \overline{x}, y, z) \Rightarrow \bigvee_{1 \leq i \leq n} (\overline{x} = \overline{x}_i \wedge y = y_i \wedge z = z_i)))$.

Thus, the formulae $\exists X.\varphi$ and $\forall X.\varphi$ can be expressed by the predicates $\Delta(f, \overline{x}, y)$ and $\ddot{\Delta}(f, \overline{x}, y, z)$. The predicates $\Delta(f, \overline{x}, y)$ and $\ddot{\Delta}(f, \overline{x}, y, z)$ can be replaced by using membership $\in$, less-than-or-equal $\leq$, $\text{fin}(\Delta)$, $\text{fin}(\ddot{\Delta})$ and other first-order logic formulae, in which $\text{fin}(\Delta)$ and $\text{fin}(\ddot{\Delta})$ can again be replaced by using membership $\in$ and less-than-or-equal $\leq$ predicates. It means that, due to the finiteness of update sets and multisets, we actually still stay with the interpretation of the first-order logic in spite of the use of second-order formulae in the syntax of $\mathbb{L}_M^{\mathfrak{L}}$.

5.2.3 Non-determinism

Non-deterministic transitions manifest themselves as a challenging task in the logical formalisation for ASMs. In [47] Stärk and Nanchen analysed various possible problems encountered in the approaches they tried by taking non-determinism into consideration, and they stated:

> "Unfortunately, the formalisation of consistency cannot be applied directly to non-deterministic ASMs. The formula $\text{Con}(R)$ (as defined in Sect. 8.1.2) expresses the property that the *union of all possible* update sets of R in a given state is consistent. This is clearly not what is meant by consistency. Therefore, in a logic for ASMs with **choose** one had to add $\text{Con}(R)$ as an atomic formula to the logic."

This statement is not universally true. For update sets (or multisets) that contain only finite updates, they can be made explicit in the formulae of a logic which captures non-deterministic transitions. Then, the formalisation of consistency as defined in Sect. 8.1.2 of [47] can still be applied to such an explicitly specified update set Δ yielded by a rule r in the form of the formula $\text{con}(r, \Delta)$, which will be defined in Subsection 5.2.4.

More precisely, our approach to the problem of non-determinism is based on the following observations:

- As a DB-ASM rule r may associate with a set $\Delta(r, S, \zeta)$ of different update sets and applying different update sets in $\Delta(r, S, \zeta)$ leads to a set of different successor states to the current state, we add the formula $[\Delta]\varphi$ into the logic to express the property that the formula φ is interpreted in the states after applying the update set Δ over the current state;

- The inclusion of the formulae $[\Delta]\varphi$, $X.\varphi$, $\forall X.\varphi$, $\text{upd}(r, \Delta)$ and $\text{upm}(r, \ddot{\Delta})$ in $\mathbb{L}_M^{\mathfrak{E}}$ empowers us to express the interpretation of a formula φ over all successor states or over some successor state after applying a DB-ASM rule r.

- The formalisation of consistency can be captured by the formula $\text{con}(r, \Delta)$ which is specific to an update set Δ yielded by a DB-ASM rule r, and is able to be extended to two versions (i.e., weak version $\text{wcon}(r)$ and strong version $\text{scon}(r)$) that describe the consistency of a rule r defined in two different senses. We will discuss the details in the next subsection.

- The underlying assumption of this approach is the finiteness of update sets and multisets, which can be assured by the definition of DB-ASMs (see Definition 3.2.2 in Chapter 3).

5.2.4 Consistency

In [47] Stärk and Nanchen used a predicate $\text{Con}(r)$ as an abbreviation for the statement that a rule r is defined and consistent. As a rule r in their work was considered to be deterministic, there was no ambiguity with the reference to the update set associated with r, i.e., a defined rule r is consistent iff the update set generated by r is consistent. However, in the case of the logic for DB-ASMs, the presence of non-determinism makes the situation a bit different.

Let r be a DB-ASM rule and Δ be an update set. Then $\text{con}(r, \Delta)$ is an abbreviation of the following formula, representing that the update set Δ generated by the rule r is consistent.

$$\text{con}(r, \Delta) \equiv \text{upd}(r, \Delta) \wedge \bigwedge_{f \in \mathbf{F}_{dyn}} \forall \overline{x}, y, y'.(\Delta(f, \overline{x}, y) \wedge \Delta(f, \overline{x}, y') \Rightarrow y = y')$$

From the above expression, it is clear that there is no any connection between the formulae $\text{con}(r, \Delta)$ and $\text{def}(r)$. That is, if $\text{con}(r, \Delta)$ is interpreted as true, it only means that Δ is an update set that can be (but not necessarily) yielded by the rule r. Since the rule r may be non-deterministic, it is possible that rule r may yield an update set Δ in one case and may not terminate in another case.

In the formalisation of the logic for DB-ASMs, sometimes it is more convenient to use the formula con(Δ), which is the abbreviation defined by

$$\text{con}(\Delta) \equiv \bigwedge_{f \in \mathbf{F}_{dyn}} \forall \overline{x}, y, y'.(\Delta(f, \overline{x}, y) \wedge \Delta(f, \overline{x}, y') \Rightarrow y = y')$$

In terms of the consistency of DB-ASM rules, there are two versions to be developed.

(1) A rule r is said to be *weakly consistent* (denoted as wcon(r)) iff r is defined and at least one update set generated by r is consistent such that

- $\text{wcon}(r) \equiv \text{def}(r) \wedge \exists \Delta.\text{con}(r, \Delta)$.

(2) A rule r is said to be *strongly consistent* (denoted as scon(r)) iff r is defined and every update set generated by r is consistent such that

- $\text{scon}(r) \equiv \text{def}(r) \wedge \forall \Delta.(\text{upd}(r, \Delta) \Rightarrow \text{con}(r, \Delta))$.

In the case that a rule r is deterministic, the weak notion of consistency coincides with the strong notion of consistency, i.e., $\text{wcon}(r) \Leftrightarrow \text{scon}(r)$. Clearly, if a rule r is not defined, then it is neither weakly nor strongly consistent.

5.2.5 Definedness

Recall that, a DB-ASM rule r is said to be defined in a state S under a variable assignment ζ (i.e., $[\![\text{def}(r)]\!]_{S,\zeta} = \textit{true}$) if the rule r can generate at least one update set which may or may not be consistent.

The properties of predicate def(r) are presented in Figure 5.1. Since the rules of DB-ASMs exclude call rules of ASMs, there is no iteration involved in each computation. Nevertheless, we have to take care of the additional effect on the definedness of a rule caused by the possible non-determinism. This has been embodied in the definedness of the choice and sequence rules in Axioms **D5** and **D6**, respectively.

- For Axiom **D5**, it might be appealing to be formalised as follows:

$$\text{def}(\textbf{choose } \overline{z} \textbf{ with } \varphi \textbf{ do } r \textbf{ enddo}) \Leftrightarrow \exists \overline{z}.(\varphi \wedge \text{def}(r))$$

By this formalisation, a choice rule would not be defined if there does not exist any value $\overline{a}$ for $\overline{z}$ such that the interpretation of $\varphi[\overline{a}/\overline{z}]$ is true. Clearly, this violates the definition of definedness for a rule since in this case a choice rule still yields an empty update set. Therefore, Axiom **D5** in Figure 5.1 takes care of this case by adding a disjunctive condition $\forall\overline{z}.(\neg\varphi)$ to the above formalisation.

- For Axiom **D6**, it should not be formulated as

$$\text{def}(\textbf{seq}\ r_1\ \ r_2\ \textbf{endseq}) \Leftrightarrow \exists\Delta.\text{def}(r_1)\wedge\text{upd}(r_1,\Delta)\wedge[\Delta]\text{def}(r_2).$$

Suppose that the rule r_1 yields two update sets Δ_1 and Δ_2, Δ_1 is inconsistent and Δ_2 is consistent. Furthermore, the rule r_2 is not defined in the successor state that is obtained by applying Δ_2 over the current state. In this case, by the above formalisation, the sequence rule **seq** r_1 r_2 **endseq** would be interpreted as being defined because Δ_1 is inconsistent and $[\Delta_1]\text{def}(r_2)$ is thus interpreted as being true. For this reason, the formulation we propose for Axiom **D6** becomes def(**seq** r_1 r_2 **endseq**) $\Leftrightarrow \text{def}(r_1)\wedge(\text{wcon}(r_1)\Rightarrow\langle r_1\rangle\text{def}(r_2))$. It says that either all update sets yielded by rule r_1 are inconsistent, or the rule r_2 is defined in at least one of the successor states after applying consistent update sets yielded by rule r_1 over the current state.

5.2.6 Update Sets

Before presenting the axioms for update sets and multisets generated by DB-ASM rules, we introduce two abbreviations $\text{inv}(\Delta,f,\overline{x})$ and $\text{inv}(\ddot{\Delta},f,\overline{x})$ asserting that the update set Δ and multiset $\ddot{\Delta}$ do not have any update to the location of the function symbol f at the argument $\overline{x}$, respectively, such that

- $\text{inv}(\Delta,f,\overline{x})\equiv\forall y.\neg\Delta(f,\overline{x},y)$.
- $\text{inv}(\ddot{\Delta},f,\overline{x})\equiv\forall y,z.\neg\ddot{\Delta}(f,\overline{x},y,z)$.

In association with a DB-ASM rule r, the properties of the predicate $\text{upd}(r,\Delta)$ are presented by Axioms **U1**-**U7** in Figure 5.2.

Axiom **U1** says that the update set yielded by an assignment rule $f(\overline{t}):=t_0$ contains exactly one update $(f,\overline{t},t_0)$. Axiom **U2** asserts that, if the formula φ is true in a conditional

D1. $\text{def}(f(\bar{t}) := t_0)$

D2. $\text{def}(\textbf{if } \varphi \textbf{ then } r \textbf{ endif}) \Leftrightarrow \neg\varphi \vee (\varphi \wedge \text{def}(r))$

D3. $\text{def}(\textbf{forall } \overline{z} \textbf{ with } \varphi \textbf{ do } r \textbf{ enddo}) \Leftrightarrow \forall \overline{z}.(\varphi \Rightarrow \text{def}(r))$

D4. $\text{def}(\textbf{par } r_1 ... r_n \textbf{ endpar}) \Leftrightarrow \text{def}(r_1) \wedge \cdots \wedge \text{def}(r_n)$

D5. $\text{def}(\textbf{choose } \overline{z} \textbf{ with } \varphi \textbf{ do } r \textbf{ enddo}) \Leftrightarrow \exists \overline{z}.(\varphi \wedge \text{def}(r)) \vee \forall \overline{z}.(\neg\varphi)$

D6. $\text{def}(\textbf{seq } r_1 \ r_2 \textbf{ endseq}) \Leftrightarrow \text{def}(r_1) \wedge (\text{wcon}(r_1) \Rightarrow \langle r_1 \rangle \text{def}(r_2))$

D7. $\text{def}(\textbf{let } \theta(t) = \rho \textbf{ in } r \textbf{ endlet}) \Leftrightarrow \text{def}(r)$

Figure 5.1: Axioms for predicate def(r)

rule **if** φ **then** r **endif**, then the update set yielded by the conditional rule is also an update set yielded by rule r. Otherwise, the conditional rule yields only an empty update set Δ.

Axiom **U3** states that the update set yielded by a forall rule **forall** $\overline{z}$ **with** φ **do** r **enddo** is the union of a set of update sets where rule r under each distinct value $\overline{a}$ for $\overline{z}$ such that the interpretation of $\varphi[\overline{a}/\overline{z}]$ is true yields exactly one update set. Similarly, Axiom **U4** states that the update set yielded by a parallel rule **par** $r_1 \dots r_n$ **endpar** is the union of a set of update sets yielded by rules $r_1, ..., r_n$ where each rule r_i ($i \in [1, n]$) yields exactly one update set. As a DB-ASM rule may be non-deterministic, a straightforward extension from the formalisation of the forall and parallel rules used in the logic for ASMs [134] would not work for Axioms **U3** and **U4**. For example, the following formalisation for a forall rule indeed asserts that the update set yielded by a forall rule is the union of all possible update sets yielded by the rule r under all values $\overline{a}$ for $\overline{z}$ such that the interpretation of $\varphi[\overline{a}/\overline{z}]$ is true.

$$\begin{aligned}&\text{upd}(\textbf{forall } \overline{z} \textbf{ with } \varphi \textbf{ do } r \textbf{ enddo}, \Delta) \Leftrightarrow \text{def}(\textbf{forall } \overline{z} \textbf{ with } \varphi \textbf{ do } r \textbf{ enddo}) \\ &\quad \wedge \bigwedge_{f \in \mathbf{F}_{dyn}} \forall \overline{x}, y.(\Delta(f, \overline{x}, y) \Leftrightarrow \exists \overline{z}.(\varphi \wedge \exists \Delta'.(\text{upd}(r, \Delta') \wedge \Delta'(f, \overline{x}, y))))\end{aligned}$$

U1. $\text{upd}(f(\bar{t}) := t_0, \Delta) \Leftrightarrow \Delta(f, \bar{t}, t_0) \wedge \forall \bar{x}, y.(\bar{x} \neq \bar{t} \vee y \neq t_0 \Rightarrow \neg\Delta(f, \bar{x}, y)) \wedge \bigwedge_{f \neq f' \wedge f \in \mathbb{F}_{dyn} \wedge f' \in \mathbb{F}_{dyn},} \forall \bar{x}, y. \neg\Delta(f', \bar{x}, y)$

U2. $\text{upd}(\textbf{if } \varphi \textbf{ then } r \textbf{ endif}, \Delta) \Leftrightarrow (\varphi \wedge \text{upd}(r, \Delta)) \vee (\neg\varphi \wedge \bigwedge_{f \in \mathbb{F}_{dyn}} \forall \bar{x}, y. \neg\Delta(f, \bar{x}, y))$

U3. $\text{upd}(\textbf{forall } \bar{z} \textbf{ with } \varphi \textbf{ do } r \textbf{ enddo}, \Delta) \Leftrightarrow \text{def}(\textbf{forall } \bar{z} \textbf{ with } \varphi \textbf{ do } r \textbf{ enddo}) \wedge \Delta = \bigcup_{\bar{z}}(\Delta' | \varphi \wedge \text{upd}(r, \Delta'))$

U4. $\text{upd}(\textbf{par } r_1 \ldots r_n \textbf{ endpar}, \Delta) \Leftrightarrow \text{def}(\textbf{par } r_1 \ldots r_n \textbf{ endpar}) \wedge \bigwedge_{i=1}^{n} \exists \Delta_i.\text{upd}(r_i, \Delta_i) \wedge \bigwedge_{f \in \mathbb{F}_{dyn}} \forall \bar{x}, y.(\Delta(f, \bar{x}, y) \Leftrightarrow \bigvee_{i=1}^{n} \Delta_i(f, \bar{x}, y))$

U5. $\text{upd}(\textbf{choose } \bar{z} \textbf{ with } \varphi \textbf{ do } r \textbf{ enddo}, \Delta) \Leftrightarrow \text{def}(\textbf{choose } \bar{z} \textbf{ with } \varphi \textbf{ do } r \textbf{ enddo}) \wedge \exists \bar{z}.(\varphi \wedge \text{upd}(r, \Delta)) \vee (\forall \bar{z}. \neg\varphi \wedge \bigwedge_{f \in \mathbb{F}_{dyn}} \forall \bar{x}, y. \neg\Delta(f, \bar{x}, y))$

U6. $\text{upd}(\textbf{seq } r_1 \; r_2 \textbf{ endseq}, \Delta) \Leftrightarrow \exists \Delta_1, \Delta_2.\text{upd}(r_1, \Delta_1) \wedge [\Delta_1]\text{upd}(r_2, \Delta_2) \wedge \bigwedge_{f \in \mathbb{F}_{dyn}} \forall \bar{x}, y.(\Delta(f, \bar{x}, y) \Leftrightarrow (\Delta_1(f, \bar{x}, y) \wedge ([\Delta_1]\text{def}(r_2) \wedge \text{inv}(\Delta_2, f, \bar{x}))) \vee (\text{con}(\Delta_1) \wedge [\Delta_1]\Delta_2(f, \bar{x}, y)))$

U7. $\text{upd}(\textbf{let } \theta(t) = \rho \textbf{ in } r \textbf{ endlet}, \Delta) \Leftrightarrow \exists \ddot{\Delta}.\text{upm}(r, \ddot{\Delta}) \wedge \bigwedge_{f \in \mathbb{F}_{dyn}} \forall \bar{x}, y.(\Delta(f, \bar{x}, y) \Leftrightarrow (f(\bar{x}) \neq t \wedge \exists z.\ddot{\Delta}(f, \bar{x}, y, z)) \vee (f(\bar{x}) = t \wedge y = \rho_{y'}(y' | \exists z.\ddot{\Delta}(f, \bar{x}, y', z))))$

Figure 5.2: Axioms for predicate upd(r,Δ)

Example 5.2.1. Let us consider a forall rule **forall** $\overline{z}$ **with** φ **do** r **enddo** which has two distinct values $\overline{a}_1$ and $\overline{a}_2$ for $\overline{z}$ such that both $\varphi[\overline{a}_1/\overline{z}]$ and $\varphi[\overline{a}_2/\overline{z}]$ are interpreted to be true. Assume that we have

- $r[\overline{a}_1/\overline{z}]$ is associated with two possible update sets $\{\Delta_{11}, \Delta_{12}\}$, and
- $r[\overline{a}_2/\overline{z}]$ is associated with two possible update sets $\{\Delta_{21}, \Delta_{22}\}$.

By the above incorrect axiom, it states that exactly one update set $\{\Delta_{11} \cup \Delta_{12} \cup \Delta_{21} \cup \Delta_{22}\}$ is associated with the forall rule. Clearly, it is not right.

Instead, an update set for the forall rule should be the union of one update set from $\{\Delta_{11}, \Delta_{12}\}$ and another update set from $\{\Delta_{21}, \Delta_{22}\}$. In this case, the forall rule should be associated with the following set of update sets:

$$\{\{\Delta_{11} \cup \Delta_{21}\}, \{\Delta_{11} \cup \Delta_{22}\}, \{\Delta_{12} \cup \Delta_{21}\}, \{\Delta_{12} \cup \Delta_{22}\}\}.$$

□

Axiom **U5** asserts that an update set yielded by a choice rule **choose** $\overline{z}$ **with** φ **do** r **enddo** is either an update set yielded by the rule r under a value $\overline{a}$ for the variable $\overline{z}$ such that $\varphi[\overline{a}/\overline{z}]$ is interpreted as true, or empty if the formula φ cannot be interpreted to be true by any values for $\overline{z}$.

Axiom **U6** asserts that an update set yielded by a sequence rule **seq** r_1 r_2 **endseq** has updates which are either in the update set Δ_1 yielded by rule r_1 in a state S and not in the update set Δ_2 yielded by the defined rule r_2 executing over the successor state $S + \Delta_1$, or in Δ_2 if Δ_1 is consistent. The following formalisation for Axiom **U6** would be incorrect.

$$\begin{aligned}\text{upd}(\textbf{seq}\ r_1\ r_2\ \textbf{endseq}, \Delta) \Leftrightarrow \exists \Delta_1.\text{upd}(r_1, \Delta_1) \wedge \bigwedge_{f \in \mathbf{F}_{dyn}} \forall \overline{x}, y.(\Delta(f, \overline{x}, y) \Leftrightarrow \\ (\Delta_1(f, \overline{x}, y) \wedge \forall \Delta_2.([\Delta_1]\text{upd}(r_2, \Delta_2) \Rightarrow \text{inv}(\Delta_2, f, \overline{x}))) \\ \vee(\text{con}(\Delta_1) \wedge \exists \Delta_2([\Delta_1]\text{upd}(r_2, \Delta_2) \wedge \Delta_2(f, \overline{x}, y))))\end{aligned}$$

The following example illustrates the problem of such a formalisation.

Example 5.2.2. Let S be a state, and r_1 and r_2 be two DB-ASM rules. Assume that the rule r_1 is executed over the state S, yielding only one update set $\Delta_1 = \{(f, \overline{t}_3, t_3)\}$ (i.e., r_1 is deterministic). Thus, in the successor state $S_1 = S + \Delta_1$ of S, the location $f(\overline{t}_3)$ has the value t_3. Now we continue to assume that the rule r_2 executing over state S_1 is associated with a set $\{\Delta_2^1, \Delta_2^2, \Delta_2^3\}$ of update sets such that

- $\Delta_2^1 = \{(f, \bar{t}_1, t_1)\}$,
- $\Delta_2^2 = \{(f, \bar{t}_2, t_2)\}$,
- $\Delta_2^3 = \{(f, \bar{t}_3, t_3)\}$.

Therefore, it is obvious to see that the rule **seq** r_1 r_2 **endseq** executing over the state S should be associated with a set $\{\Delta^1, \Delta^2, \Delta^3\}$ of update sets such that

- $\Delta^1 = \{(f, \bar{t}_1, t_1), (f, \bar{t}_3, t_3)\}$,
- $\Delta^2 = \{(f, \bar{t}_2, t_2), (f, \bar{t}_3, t_3)\}$,
- $\Delta^3 = \{(f, \bar{t}_3, t_3)\}$.

However, the above formalisation asserts that the following update sets $\{(f, \bar{t}_1, t_1), (f, \bar{t}_2, t_2), (f, \bar{t}_3, t_3)\}$ and $\{(f, \bar{t}_1, t_1), (f, \bar{t}_2, t_2)\}$ could also be two associated update sets of the rule **seq** r_1 r_2 **endseq** executing over the state S. □

Axiom **U7** asserts that an update multiset may be collapsed into an update set by aggregating the update values of locations if these locations have been assigned a location operator by a let rule, or ignoring the multiplicity of updates if their locations have no assigned location operators.

5.2.7 Update Multisets

As formalised in Definition 3.2.1, a DB-ASM rule is also associated with a set of update multisets. Axioms **UM1-UM7** in Figure 5.3 assert how an update multiset is yielded by a DB-ASM rule. Basically, the axioms for predicate $\text{upm}(r, \ddot{\Delta})$ are defined in a similar way to the axioms for predicate $\text{upd}(r, \Delta)$, except for Axioms **UM6** and **UM7**. So we explain these two axioms in particular.

Axiom **UM6** asserts that, for each update multiset $\ddot{\Delta}$ yielded by a sequence rule **seq** r_1 r_2 **endseq**, there exists an update multiset $\ddot{\Delta}_1$ yielded by the rule r_1 over the current state S and an update multiset $\ddot{\Delta}_2$ yielded by the rule r_2 over the successor state $S + \Delta_1$, where the update set Δ_1 is obtained by ignoring the multiplicity of all updates in $\ddot{\Delta}_1$. Furthermore, updates to each location $f(\overline{x})$ in $\ddot{\Delta}$ are either from $\ddot{\Delta}_1$ when there are no updates to $f(\overline{x})$ in update multiset $\ddot{\Delta}_2$, or from $\ddot{\Delta}_2$ when Δ_1 is consistent.

Axiom **UM7** asserts that an update multiset $\ddot{\Delta}$ yielded by a let rule **let** $\theta(t) = \rho$ **in** r **endlet** contains updates that are either aggregated from update values to the location t

in the update multiset $\ddot{\Delta}'$ yielded by rule r or exactly the same as updates in $\ddot{\Delta}'$ if their locations have no assigned location operators. In contrast to Axiom **U7**, when a location operator is not assigned to a location by the let rule, the multiplicity of updates will remain in $\ddot{\Delta}$.

Note that, as formalised in Definition 5.2.2, an update $(f(\bar{t}), t_0)$ with the multiplicity n in an update multiset $\ddot{\Delta}$ implies that the formulae $\ddot{\Delta}(f, \bar{t}, t_0, 0), ..., \ddot{\Delta}(f, \bar{t}, t_0, n-1)$ representing the occurred elements are evaluated to be *true*. To obtain the aggregated value t' of an update $(f(\bar{t}), t')$ in an update set, which comes from multiple updates $(f(\bar{t}), t_1'), ..., (f(\bar{t}), t_n')$ in an update multiset $\ddot{\Delta}$, we need a ρ-term $\rho_y(y|\exists z.\ddot{\Delta}(f, \bar{x}, y, z))$ such that $val_{S,\zeta[\bar{x}\mapsto\bar{t}]}(\rho_y(y|\exists z.\ddot{\Delta}(f, \bar{x}, y, z))) = \rho(\{\{t_1', ..., t_n'\}\}) = t'$. Therefore, ρ-terms play a vital role in formalising operations over a multiset.

Remark 5.2.2. The inclusion of the parameters Δ and $\ddot{\Delta}$ in the predicates $\text{upd}(r, \Delta)$ and $\text{upm}(r, \ddot{\Delta})$ is important because a DB-ASM rule r may associate with multiple update sets or multisets and we need a way to specify which update set or multiset yielded by rule r is of our interest. Furthermore, the use of predicates $\text{upd}(r, \Delta)$ and $\text{upm}(r, \ddot{\Delta})$ provides the capability to specify the interrelationship among update sets or multisets associated with possibly different rules, e.g., Axioms **U6**, **UM6** and **UM7**.

5.3 A Proof System

With the use of modal operator [ ] for an update set Δ (i.e., $[\Delta]$), the logic for DB-ASMs becomes a multi-modal logic. In doing so, we may use it to reason about properties of a database transformation over different states of a run. Semantically, formulae of $\mathbb{L}_M^{\mathfrak{L}}$ are interpreted in states that are represented as a Kripke frame.

Definition 5.3.1. A *Kripke frame* is a pair $(U_{\mathfrak{K}}, R_{\mathfrak{K}})$ consisting of

- a universe $U_{\mathfrak{K}}$ that is a non-empty set of states, and
- a binary accessibility relation $R_{\mathfrak{K}}$ on $U_{\mathfrak{K}}$ such that $(S, S') \in R_{\mathfrak{K}}$ for $S, S' \in U_{\mathfrak{K}}$.

Straightforwardly, a database transformation T can be regarded as a Kripke frame $(U_{\mathfrak{K}}, R_{\mathfrak{K}})$ where a non-empty set $\mathcal{S}_T$ of states is in $U_{\mathfrak{K}}$ and a one-step transition relation δ_T corresponds to $R_{\mathfrak{K}}$.

Before presenting the axioms and inference rules of a proof system for $\mathbb{L}_M^{\mathfrak{L}}$, we first define the notions of implication and derivability.

UM1. $\mathrm{upm}(f(\bar{t}) := t_0, \ddot{\Delta}) \Leftrightarrow \ddot{\Delta}(f, \bar{t}, t_0, 0) \wedge \forall \overline{x}, y, z.(\ddot{\Delta}(f, \overline{x}, y, z) \Rightarrow \overline{x} = \bar{t} \wedge y = t_0 \wedge$
$z = 0) \wedge \bigwedge_{f \neq f' \wedge f \in \mathbf{F}_{dyn} \wedge f' \in \mathbf{F}_{dyn}} \forall \overline{x}, y, z. \neg \ddot{\Delta}(f', \overline{x}, y, z)$

UM2. $\mathrm{upm}(\textbf{if } \varphi \textbf{ then } r \textbf{ endif}, \ddot{\Delta}) \Leftrightarrow (\varphi \wedge \mathrm{upm}(r, \ddot{\Delta})) \vee$
$(\neg \varphi \wedge \bigwedge_{f \in \mathbf{F}_{dyn}} \forall \overline{x}, y, z. \neg \ddot{\Delta}(f, \overline{x}, y, z))$

UM3. $\mathrm{upm}(\textbf{forall } \overline{z} \textbf{ with } \varphi \textbf{ do } r \textbf{ enddo}, \ddot{\Delta}) \Leftrightarrow \mathrm{def}(\textbf{forall } \overline{z} \textbf{ with } \varphi \textbf{ do } r \textbf{ enddo}) \wedge$
$\ddot{\Delta} = \biguplus_{\overline{z}}(\ddot{\Delta}' | \varphi \wedge \mathrm{upm}(r, \ddot{\Delta}'))$

UM4. $\mathrm{upm}(\textbf{par } r_1 \ldots r_n \textbf{ endpar}, \ddot{\Delta}) \Leftrightarrow \mathrm{def}(\textbf{par } r_1 \ldots r_n \textbf{ endpar}) \wedge \bigwedge_{i=1}^{n} \exists \ddot{\Delta}_i . \mathrm{upm}(r_i, \ddot{\Delta}_i) \wedge$
$\bigwedge_{f \in \mathbf{F}_{dyn}} \forall \overline{x}, y, w'.(\ddot{\Delta}(f, \overline{x}, y, w') \Leftrightarrow w' < \sum_{w^i_{max}} (w^i_{max} + 1 | \bigvee_{i=1}^{n} \ddot{\Delta}_i(f, \overline{x}, y, w^i_{max})$
$\wedge \forall w.(\ddot{\Delta}_i(f, \overline{x}, y, w) \Rightarrow w \leq w^i_{max})))$

UM5. $\mathrm{upm}(\textbf{choose } \overline{z} \textbf{ with } \varphi \textbf{ do } r \textbf{ enddo}, \ddot{\Delta}) \Leftrightarrow \exists \overline{z}.(\varphi \wedge \mathrm{upm}(r, \ddot{\Delta}))$
$\vee (\forall \overline{z}. \neg \varphi \wedge \bigwedge_{f \in \mathbf{F}_{dyn}} \forall \overline{x}, y, z. \neg \ddot{\Delta}(f, \overline{x}, y, z))$

UM6. $\mathrm{upm}(\textbf{seq } r_1 \; r_2 \textbf{ endseq}, \ddot{\Delta}) \Leftrightarrow \exists \ddot{\Delta}_1, \ddot{\Delta}_2, \Delta_1 . \mathrm{upm}(r_1, \ddot{\Delta}_1) \wedge \Delta_1 = AsSet(\ddot{\Delta}_1) \wedge$
$[\Delta_1]\mathrm{upm}(r_2, \ddot{\Delta}_2) \wedge \bigwedge_{f \in \mathbf{F}_{dyn}} \forall \overline{x}, y, w.(\ddot{\Delta}(f, \overline{x}, y, w) \Leftrightarrow (\ddot{\Delta}_1(f, \overline{x}, y, w) \wedge$
$([\Delta_1]\mathrm{def}(r_2) \wedge \mathrm{inv}(\ddot{\Delta}_2, f, \overline{x}))) \vee (\mathrm{con}(\Delta_1) \wedge [\Delta_1]\ddot{\Delta}_2(f, \overline{x}, y, w)))$

UM7. $\mathrm{upm}(\textbf{let } \theta(t) = \rho \textbf{ in } r \textbf{ endlet}, \ddot{\Delta}) \Leftrightarrow \exists \ddot{\Delta}' . \mathrm{upm}(r, \ddot{\Delta}') \wedge$
$\bigwedge_{f \in \mathbf{F}_{dyn}} \forall \overline{x}, y, z.(\ddot{\Delta}(f, \overline{x}, y, z) \Leftrightarrow (f(\overline{x}) \neq t \wedge \ddot{\Delta}'(f, \overline{x}, y, z))$
$\vee (f(\overline{x}) = t \wedge y = \rho_{y'}(y' | \exists z. \ddot{\Delta}'(f, \overline{x}, y', z))))$

Figure 5.3: Axioms for predicate upm(r,$\ddot{\Delta}$)

Definition 5.3.2. Let M be a DB-ASM. Then a formula φ is said to be *implied* by a set Ψ of formulae with respect to M (denoted as $\Psi \models_M \varphi$) if for all states S and variable assignments ζ of M, the following condition is satisfied.

- If $[\![\psi]\!]_{S,\zeta} = true$ for every $\psi \in \Psi$, then $[\![\varphi]\!]_{S,\zeta} = true$.

A formula φ is called *valid* in M if $[\![\varphi]\!]_{S,\zeta} = true$ for all states S and variable assignments ζ of M.

Definition 5.3.3. Let M be a DB-ASM. Then a formula φ is said to be *derived* from a set Ψ of formulae with respect to M (denoted as $\Psi \vdash_M \varphi$) if φ is derivable from Ψ by successively applying the axioms and inference rules of the proof system defined in Subsection 5.3.1.

We also need to define equivalence between two DB-ASM rules.

Definition 5.3.4. Let r_1 and r_2 be two DB-ASM rules. Then r_1 and r_2 are *equivalent* (denoted as $r_1 \simeq r_2$) if, for all states and variable assignments, the following condition is satisfied:

$$\forall\Delta.(\text{upd}(r_1, \Delta) \Leftrightarrow \text{upd}(r_2, \Delta)).$$

By the above definition, the fact that two rules r_1 and r_2 are either both defined or both undefined can be derived. That is,

$$r_1 \simeq r_2 \Rightarrow (\text{def}(r_1) \Leftrightarrow \text{def}(r_2)).$$

As a straightforward consequence of Definition 5.3.4, the interpretation on weak and strong consistencies of two equivalent rules r_1 and r_2 coincides. That is,

- wcon(r_1)=wcon(r_2), and
- scon(r_1)=scon(r_2).

5.3.1 Axioms and Inference Rules

We present a set of axioms and inference rules that constitute a proof system of the logic for DB-ASMs.

- The following axioms assert the properties of def(r).

D1.-D7. in Figure 5.1

- The following axioms assert the properties of $\mathrm{upd}(r, \Delta)$.

  U1.-U7. in Figure 5.2

- The following axioms assert the properties of $\mathrm{upm}(r, \ddot{\Delta})$.

  UM1.-UM7. in Figure 5.3

- Axiom **M1** and Rules **M2-M3** are from the axiom system K of modal logic, which is the weakest normal modal logic system [95]. More specifically, Axiom **M1** is called *Distribution Axiom* of K and Rule **M2** is called *Necessitation Rule* of K. Rule **M3** is the inference rule called *Modus Ponens* in classical logic. By using these axiom and rules together, we are able to derive all modal properties that are valid in Kripke frames.

  M1. $[\Delta](\varphi \Rightarrow \psi) \Rightarrow [\Delta]\varphi \Rightarrow [\Delta]\psi$

  M2. $\varphi \vdash [\Delta]\varphi$

  M3. $\varphi, \varphi \Rightarrow \psi \vdash \psi$

- Axiom **M4** asserts that, if an update set Δ is not consistent, then there is no successor state after applying Δ over the current state and thus $[\Delta]\varphi$ is interpreted as true for any formula φ. Since applying an update set Δ over the current state is deterministic, Axiom **M5** describes the deterministic accessibility relation in terms of $[\Delta]$.

  M4. $\neg\mathrm{con}(\Delta) \Rightarrow [\Delta]\varphi$

  M5. $\neg[\Delta]\varphi \Leftrightarrow [\Delta]\neg\varphi$

- Axiom **M6** is also called *Barcan Axiom*, saying that all states in a run of a database transformation have the same base set, and thus quantifiers in all states always range over the same set of elements.

  M6. $\forall x.[\Delta]\varphi \Rightarrow [\Delta]\forall x.\varphi$

- Axioms **M7** and **M8** assert that the interpretation of static and pure formulae is the same in all states of a database transformation, which is not affected by the execution of any DB-ASM rule r. Note that, depending on the logic that is parameterised into the logic of meta-finite states, static and pure formulae might not be first-order formulae.

  M7. $\text{con}(r, \Delta) \wedge \varphi \Rightarrow [\Delta]\varphi$ for static and pure φ

  M8. $\text{con}(r, \Delta) \wedge [\Delta]\varphi \Rightarrow \varphi$ for static and pure φ

- Axiom **A1** asserts that, if a consistent update set Δ does not contain any update to the location $f(\overline{x})$, then the content of the location $f(\overline{x})$ in a successor state after applying the update set Δ is the same as its content in the current state. Axiom **A2** asserts that, if a consistent update set Δ does contain an update which updates the content of the location $f(\overline{x})$ to y, then the content of the location $f(\overline{x})$ in the successor state after applying the update set Δ is equal to y. Axiom **A3** says that, if a DB-ASM rule r yields an update set Δ, then the rule r is defined. Axiom **A4** says that, if a DB-ASM rule r yields an update multiset, then the rule r also yields an update set.

  A1. $\text{con}(\Delta) \wedge \text{inv}(\Delta, f, \overline{x}) \wedge f(\overline{x}) = y \Rightarrow [\Delta]f(\overline{x}) = y$

  A2. $\text{con}(\Delta) \wedge \Delta(f, \overline{x}, y) \Rightarrow [\Delta]f(\overline{x}) = y$

  A3. $\text{upd}(r, \Delta) \Rightarrow \text{def}(r)$

  A4. $\text{upm}(r, \ddot{\Delta}) \Rightarrow \exists \Delta.\text{upd}(r, \Delta)$

- The following are axiom schemes from classical logic.

  P1. $\varphi \Rightarrow (\psi \Rightarrow \varphi)$

  P2. $(\varphi \Rightarrow (\psi \Rightarrow \chi)) \Rightarrow ((\varphi \Rightarrow \psi) \Rightarrow (\varphi \Rightarrow \chi))$

  P3. $(\neg\varphi \Rightarrow \neg\psi) \Rightarrow (\psi \Rightarrow \varphi)$

- Axiom **EG**, called *Existential Generalisation* in classical logic, says that some static term t satisfying a pure formula φ implies that there exists some element satisfying φ. Axiom **UI**, called *Universal Instantiation* in classical logic, says that a pure formula

φ satisfied by all elements implies that there exists some static term t satisfying φ. Note that, all the variables in $\mathbb{L}_M^{\mathfrak{L}}$ are restricted to range only over elements in the database part of a state, which are finite.

EG. $\varphi[t/x] \Rightarrow \exists x.\varphi$ if φ is pure, t is static and x ranges over B_{db}

UI. $\forall x.\varphi \Rightarrow \varphi[t/x]$ if φ is pure, t is static and x ranges over B_{db}

- The following are the equality axioms from the first-order logic with equality. Axiom **EQ1** asserts the reflexivity property, Axiom **EQ2** asserts the substitutions for functions, Axiom **EQ3** asserts the substitutions for predicates and Axiom **EQ4** asserts the substitutions for ρ-terms. Again, terms occurring in the axioms are restricted to be static, which do not contain any dynamic function symbols.

EQ1. $t = t$ for static term t

EQ2. $t_1 = t_{n+1} \wedge ... \wedge t_n = t_{2n} \Rightarrow f(t_1, ..., t_n) = f(t_{n+1}, ..., t_{2n})$ for any function f and static terms t_i $(i = 1, ..., 2n)$

EQ3. $t_1 = t_{n+1} \wedge ... \wedge t_n = t_{2n} \Rightarrow p(t_1, ..., t_n) = p(t_{n+1}, ..., t_{2n})$ for any predicate P and static terms t_i $(i = 1, ..., 2n)$

EQ4. $t_1 = t_2 \wedge (\varphi_1(\overline{x}, \overline{y}) \Leftrightarrow \varphi_2(\overline{x}, \overline{y})) \Rightarrow \rho_{\overline{x}}(t_1|\varphi_1(\overline{x}, \overline{y})) = \rho_{\overline{x}}(t_2|\varphi_2(\overline{x}, \overline{y}))$ for pure formulae $\varphi_i(\overline{x}, \overline{y})$ and static terms t_i $(i = 1, 2)$

- The following axiom is taken from dynamic logic, asserting that executing a sequence rule equals to executing the rules inside the sequence rule sequentially.

DY1. $[\textbf{seq}\ r_1\ r_2\ \textbf{endseq}]\varphi \Leftrightarrow [r_1][r_2]\varphi$

- Axiom **E** is the extensionality axiom following Definition 5.3.4.

E. $r_1 \simeq r_2 \Rightarrow \forall\Delta.\text{upd}(r_1, \Delta) \Leftrightarrow \text{upd}(r_2, \Delta)$

5.4 Soundness

We prove the soundness of the proof system by deriving valid properties of other systems.

Lemma 5.4.1. *The following modal axioms and rules used in the logic for ASMs [47] are derivable in* $\mathbb{L}^{\mathfrak{L}}_{M}$[1].

(1) $[r](\varphi \Rightarrow \psi) \Rightarrow [r]\varphi \Rightarrow [r]\psi$

(2) $\varphi \vdash [r]\varphi$

(3) $\neg wcon(r) \Rightarrow [r]\varphi$

(4) $[r]\varphi \Leftrightarrow \neg[r]\neg\varphi$

Proof. We prove each property in the following.

- Proof for Property (1):
  - By $[r]\varphi = \forall\Delta.(\text{upd}(r,\Delta) \Rightarrow [\Delta]\varphi)$, we have $[r](\varphi \Rightarrow \psi) \wedge [r]\varphi = \forall\Delta.(\text{upd}(r,\Delta) \Rightarrow [\Delta](\varphi \Rightarrow \psi)) \wedge \forall\Delta.(\text{upd}(r,\Delta) \Rightarrow [\Delta]\varphi)$.
  - By the axioms from classical logic, we have $[r](\varphi \Rightarrow \psi) \wedge [r]\varphi = \forall\Delta.(\text{upd}(r,\Delta) \Rightarrow ([\Delta](\varphi \Rightarrow \psi) \wedge [\Delta]\varphi))$.
  - Then by Axiom **M1**: $[\Delta](\varphi \Rightarrow \psi) \Rightarrow [\Delta]\varphi \Rightarrow [\Delta]\psi$, we can get $\forall\Delta.(\text{upd}(r,\Delta) \Rightarrow ([\Delta](\varphi \Rightarrow \psi) \wedge [\Delta]\varphi)) \Rightarrow \forall\Delta.(\text{upd}(r,\Delta) \Rightarrow [\Delta]\psi)$.

  Therefore, $[r](\varphi \Rightarrow \psi) \Rightarrow [r]\varphi \Rightarrow [r]\psi$ is derivable.

- Proof for Property (2):

  As discussed before, each DB-ASM rule is defined. Thus, for a DB-ASM rule r, we assume that $\Delta(r, S, \zeta) = \{\Delta_1, ..., \Delta_n\}$.

  - By Rule **M2**: $\varphi \vdash [\Delta]\varphi$, we have $\varphi \vdash [\Delta_i]\varphi$ $(i = 1, ..., n)$ for all update sets $\{\Delta_1, .., \Delta_n\}$ generated by r, i.e., $\varphi \vdash \forall\Delta.(\text{upd}(r,\Delta) \Rightarrow [\Delta]\varphi)$.
  - By the definition that $[r]\varphi = \forall\Delta.(\text{upd}(r,\Delta) \Rightarrow [\Delta]\varphi)$, we can get $\varphi \vdash [r]\varphi$.

  Therefore, $\varphi \vdash [r]\varphi$ is derivable.

[1]Property (3) and Property (4) are valid only under the assumption that the rule r is defined and deterministic.

- Proof for Property (3):

  - By $\text{wcon}(r) \Leftrightarrow \text{def}(r) \wedge \exists\Delta.\text{con}(r,\Delta)$ as defined in Subsection 5.2.4 and the fact that $\text{def}(r)$ is trivial for a DB-ASM rule r, we have $\neg\text{wcon}(r) \Leftrightarrow \neg\exists\Delta.\text{con}(r,\Delta)$.
  - By $\text{con}(r,\Delta) \Leftrightarrow (\text{upd}(r,\Delta) \wedge \text{con}(\Delta))$, we have $\neg\text{wcon}(r) \Leftrightarrow \neg\exists\Delta.(\text{upd}(r,\Delta) \wedge \text{con}(\Delta))$.
  - Since a rule r in [134] is deterministic and a DB-ASM rule is always defined, we get $\neg\text{wcon}(r) \Leftrightarrow \neg\text{con}(\Delta)$.
  - By Axiom **M4**: $\neg\text{con}(\Delta) \Rightarrow [\Delta]\varphi$ and the fact that a defined rule r in [134] yields exactly one update set Δ, we get $\neg\text{wcon}(r) \Rightarrow [r]\varphi$.

  Therefore, $\neg\text{wcon}(r) \Rightarrow [r]\varphi$ is derivable if a rule r is defined and deterministic.

- Proof for Property (4):

  - By $[r]\varphi = \forall\Delta.(\text{upd}(r,\Delta) \Rightarrow [\Delta]\varphi)$, we have $\neg[r]\neg\varphi = \exists\Delta.(\text{upd}(r,\Delta) \wedge \neg[\Delta]\neg\varphi)$.
  - By Axiom **M5**: $\neg[\Delta]\varphi \Leftrightarrow [\Delta]\neg\varphi$, we have $\neg[r]\neg\varphi = \exists\Delta.(\text{upd}(r,\Delta) \wedge [\Delta]\varphi)$.
  - When the rule is defined and deterministic, it means that the interpretation of the formula $\forall\Delta.(\text{upd}(r,\Delta) \Rightarrow [\Delta]\varphi)$ coincides the interpretation of the formula $\exists\Delta.(\text{upd}(r,\Delta) \wedge [\Delta]\varphi)$.

  Therefore, $[r]\varphi \Leftrightarrow \neg[r]\neg\varphi$ is derivable if a rule r is defined and deterministic.

□

Remark 5.4.1. The logic for ASMs [47] is deterministic by excluding the choice rule, whereas the logic for DB-ASMs includes the choice rule. Therefore, the formula $\text{Con}(R)$ used by **Axiom 5** (i.e., $\neg\text{Con}(R) \Rightarrow [R]\varphi$) in the logic for ASMs [47] indeed corresponds to the weak version of the consistency (i.e., $\text{wcon}(r)$) in the context of the logic for DB-ASMs.

Lemma 5.4.2. *The following properties are derivable in* $\mathbb{L}^{\mathfrak{L}}_{M}$.

(5) $con(r,\Delta) \wedge [\Delta]f(\overline{x}) = y \Rightarrow \Delta(f,\overline{x},y) \vee (inv(\Delta,f,\overline{x}) \wedge f(\overline{x}) = y)$

(6) $con(r,\Delta) \wedge [\Delta]\varphi \Rightarrow \neg[\Delta]\neg\varphi$

(7) $[\Delta]\exists x.\varphi \Rightarrow \exists x.[\Delta]\varphi$

(8) $[\Delta]\varphi_1 \wedge [\Delta]\varphi_2 \Rightarrow [\Delta](\varphi_1 \wedge \varphi_2)$

Proof. Property (5) is derivable by applying Axioms **A1** and **A2**. Property (6) is a straightforward result of Axiom **M5**. Property (7) can be derived by applying Axioms **M5** and **M6**. For Property (8), it is derivable by using Axioms **M1-M3**.

□

Lemma 5.4.3. *The following properties in [77] are also derivable in* $\mathbb{L}^{\mathfrak{C}}_{M}$.

- $\overline{x} = \overline{t} \Rightarrow (y = t_0 \Leftrightarrow [f(\overline{t}) := t_0]f(\overline{x}) = y)$
- $\overline{x} \neq \overline{t} \Rightarrow (y = f(x) \Leftrightarrow [f(\overline{t}) := t_0]f(\overline{x}) = y)$

However, the principles 41 and 42 mentioned in [47] are not derivable in our logic. In DB-ASMs, two parallel computations may produce an update multiset in which there are identical updates to a location assigned with a location operator. Therefore, the following statement is not true.

$$[\textbf{par } r\ r \textbf{ endpar}]\varphi \Leftrightarrow [r]\varphi$$

Example 5.4.1. Let S be a state. Suppose that firing a DB-ASM rule r over S yields an update multiset $\ddot{\Delta}_1$=$\{\!\{(f_1, a, 3, 0), (f_1, a, 3, 1), (f_2, a, 1, 0)\}\!\}$, then firing the DB-ASM rule **par** $r\ r$ **endpar** over S would yield $\ddot{\Delta}_2 = \{\!\{(f_1, a, 3, 0), (f_1, a, 3, 1), (f_2, a, 1, 0), (f_1, a, 3, 2), (f_1, a, 3, 3), (f_2, a, 1, 1)\}\!\}$. If $\theta(f_1(a)) = \Pi$ and $\theta(f_1(a)) = \bot$ where $\Pi = (id, \times, id)$ for the binary multiplication function $\times$ and the identity function id, then

- applying rule r over S may lead to the update set $\Delta_1 = \{(f_1, a, 9), (f_2, a, 1)\}$ such that
  - $f_1^{S+\Delta_1}(a) = \Pi(3, 3) = 9$, and
  - $f_2^{S+\Delta_1}(a) = 1$.
- applying rule **par** $r\ r$ **endpar** over S may lead to the update set $\Delta_2 = \{(f_1, a, 81), (f_2, a, 1)\}$ such that
  - $f_1^{S+\Delta_2}(a) = \Pi(3, 3, 3, 3) = 81$, and

- $f_2^{S+\Delta_2}(a) = 1$.

□

Following the approach of defining the predicate joinable in [47], we define the predicate joinable over two DB-ASM rules as follows. As DB-ASM rules are allowed to be non-deterministic, the predicate joinable(r_1, r_2) means that there exists a pair of update sets without conflicting updates, which are yielded by the rules r_1 and r_2, respectively. Then, based on the use of predicate joinable, the properties in Lemma 5.4.4 are all derivable.

$$\text{joinable}(r_1, r_2) :\equiv \exists\Delta_1, \Delta_2.(\text{upd}(r_1, \Delta_1)\wedge\text{upd}(r_2, \Delta_2)\wedge \bigwedge_{f\in\mathbf{F}_{dyn}} \forall\overline{x}, y, y'.(\Delta_1(f, \overline{x}, y) \wedge \Delta_2(f, \overline{x}, y') \Rightarrow y = y'))$$

Lemma 5.4.4. *The following properties for weak consistency are derivable in* $\mathbb{L}^{\mathfrak{L}}_M$.

(9) $\text{wcon}(f(\overline{t}) := t_0)$

(10) $\text{wcon}(\textbf{if } \varphi \textbf{ then } r \textbf{ endif}) \Leftrightarrow \neg\varphi \vee (\varphi\wedge\text{wcon}(r))$

(11) $\text{wcon}(\textbf{forall } \overline{z} \textbf{ with } \varphi \textbf{ do } r \textbf{ enddo}) \Leftrightarrow \forall\overline{z}.(\varphi \Rightarrow\text{wcon}(r)\wedge \forall\overline{z}'.(\varphi[\overline{z}'/\overline{z}] \Rightarrow\text{joinable}(r, r[\overline{z}'/\overline{z}])))$

(12) $\text{wcon}(\textbf{par } r_1 \ldots r_n \textbf{ endpar}) \Leftrightarrow \text{wcon}(r_1) \wedge ...\wedge\text{wcon}(r_n) \wedge \bigwedge_{1\leq i\neq j\leq n} \text{joinable}(r_i, r_j)$

(13) $\text{wcon}(\textbf{choose } \overline{z} \textbf{ with } \varphi \textbf{ do } r \textbf{ enddo}) \Leftrightarrow \exists\overline{z}.(\varphi\wedge \text{wcon}(r)) \vee \forall\overline{z}.(\neg\varphi)$

(14) $\text{wcon}(\textbf{seq } r_1 \ r_2 \textbf{ endseq}) \Leftrightarrow \text{def}(r_1) \wedge \exists\Delta_1.\text{con}(r_1, \Delta_1) \wedge [\Delta_1]\text{wcon}(r_2)$

(15) $\text{wcon}(\textbf{let } \theta(t) = \rho \textbf{ in } r \textbf{ endlet})\Leftrightarrow \exists\Delta_1, \Delta_2.(\text{def}(r)\wedge\text{upd}(r, \Delta_1)\wedge\text{con}(\Delta_2)\wedge \bigvee_{f\in\mathbf{F}_{dyn}} \forall\overline{x}, y.(\Delta_1(f, \overline{x}, y) \wedge f(\overline{x}) \neq t \Leftrightarrow \Delta_2(f, \overline{x}, y)))$

Property (17) in Lemma 5.4.5 asserts that formula φ' is interpreted to be true after applying the rule **choose** $\overline{z}$ **with** φ **do** r **enddo** iff it is true in all successor states or it is true in the current state if there is no any successor state.

Lemma 5.4.5. *The following properties for the formula* $[r]\varphi$ *are derivable in* $\mathbb{L}^{\mathfrak{L}}_M$.

(16) $[\textbf{if } \varphi \textbf{ then } r \textbf{ endif}]\varphi' \Leftrightarrow (\varphi \wedge [r]\varphi') \vee (\neg\varphi \wedge \varphi')$

(17) [**choose** $\overline{z}$ **with** φ **do** r **enddo**]$\varphi' \Leftrightarrow \forall\overline{z}.(\varphi \Rightarrow [r]\varphi') \vee \forall\overline{z}.(\neg\varphi \Rightarrow \varphi')$

Proof. These properties can be proven based on the logical equivalence $[r]\varphi \equiv \forall\Delta.(\mathrm{upd}(r,\Delta) \Rightarrow [\Delta]\varphi)$ as follows.

- Proof for Property (16): We have [**if** φ **then** r **endif**]$\varphi' \Leftrightarrow \forall\Delta.$(upd(**if** φ **then** r **endif**,Δ) $\Rightarrow [\Delta]\varphi'$). Then by Axiom **U2**, we have [**if** φ **then** r **endif**]$\varphi' \Leftrightarrow \forall\Delta.(((\varphi\wedge\mathrm{upd}(r,\Delta))\vee(\neg\varphi\wedge \bigwedge_{f\in\mathbb{F}_{dyn}} \forall\overline{x},y.\neg\Delta(f,\overline{x},y))) \Rightarrow [\Delta]\varphi')$. By $(\varphi_1\vee\varphi_2 \Rightarrow \varphi) \Leftrightarrow ((\varphi_1 \Rightarrow \varphi) \vee (\varphi_2 \Rightarrow \varphi))$, we have [**if** φ **then** r **endif**]$\varphi' \Leftrightarrow \forall\Delta.((((\varphi\wedge\mathrm{upd}(r,\Delta)) \Rightarrow [\Delta]\varphi')) \vee ((\neg\varphi \wedge \bigwedge_{f\in\mathbb{F}_{dyn}} \forall\overline{x},y.\neg\Delta(f,\overline{x},y)) \Rightarrow [\Delta]\varphi')) \Leftrightarrow (\varphi \wedge [r]\varphi') \vee (\neg\varphi \wedge \varphi')$.

- Proof for Property (17): Using the same approach as in the proof for Property (16), we have [**choose** $\overline{z}$ **with** φ **do** r **enddo**]$\varphi' \Leftrightarrow \forall\Delta.$(upd(**choose** $\overline{z}$ **with** φ **do** r **enddo**,Δ) $\Rightarrow [\Delta]\varphi'$). Then by Axiom **U5**, we have [**choose** $\overline{z}$ **with** φ **do** r **enddo**]$\varphi' \Leftrightarrow \forall\Delta.$(def(**choose** $\overline{z}$ **with** φ **do** r **enddo**) $\wedge\exists\overline{z}.(\varphi\wedge$ upd$(r,\Delta))\vee(\forall\overline{z}.\neg\varphi\wedge \bigwedge_{f\in\mathbb{F}_{dyn}} \forall\overline{x},y.\neg\Delta(f,\overline{x},y))) \Rightarrow [\Delta]\varphi')$. Then by $(\varphi_1\vee\varphi_2 \Rightarrow \varphi) \Leftrightarrow ((\varphi_1 \Rightarrow \varphi)\vee(\varphi_2 \Rightarrow \varphi))$, we can prove that [**choose** $\overline{z}$ **with** φ **do** r **enddo**]$\varphi' \Leftrightarrow\forall\overline{z}.(\varphi \Rightarrow [r]\varphi')\vee\forall\overline{z}.(\neg\varphi \Rightarrow \varphi')$.

□

The properties in Lemma 5.4.6 state that a parallel composition is commutative and associative while a sequential composition is associative.

Lemma 5.4.6. *The following properties for parallel and sequential compositions are derivable in* $\mathbb{L}^{\mathfrak{L}}_{M}$.

(18) **par** r_1 r_2 **endpar** $\simeq$ **par** r_2 r_1 **endpar**

(19) **par** (**par** r_1 r_2 **endpar**) r_3 **endpar** $\simeq$ **par** r_1 (**par** r_2 r_3 **endpar**) **endpar**

(20) **seq** (**seq** r_1 r_2 **endseq**) r_3 **endseq** $\simeq$ **seq** r_1 (**seq** r_2 r_3 **endseq**) **endseq**

Lemma 5.4.7. *The extensionality axiom for transition rules in [134] is derivable in* $\mathbb{L}^{\mathfrak{L}}_{M}$.

(21) $r_1 \simeq r_2 \Rightarrow ([r_1]\varphi \Leftrightarrow [r_2]\varphi)$

Based on the above derivable properties, we have the following theorem for the soundness of the proof system.

Theorem 5.4.1. *Let* M *be a DB-ASM and* Φ *a set of sentences. If* $\Phi \vdash_M \varphi$, *then* $\Phi \models_M \varphi$.

5.5 Completeness

In this section we discuss the completeness of the proof system defined in the previous section. As specified in Definition 5.5.1, the logic $\mathbb{L}(\Upsilon, \mathfrak{L})$ of meta-finite states may be parameterised by any logic $\mathfrak{L}$ suitable for first-order structures. However, if a chosen logic $\mathfrak{L}$ is not complete, then the logic for DB-ASMs that is built upon the logic $\mathbb{L}(\Upsilon, \mathfrak{L})$ of meta-finite states will not be complete as well. For this reason, we choose the first-order logic which is well-known to be complete and investigate the completeness of the proof system for the logic for DB-ASMs (i.e., $\mathbb{L}_M^{\mathfrak{L}}$ for $\mathfrak{L} = FO$).

We start by embedding the first-order logic into the logic of meta-finite states. The syntax and semantics of first-order formulae are defined in a standard way.

Definition 5.5.1. The terms and formulae of the logic $\mathbb{L}(\Upsilon, FO)$ over meta-finite states with signature Υ which is parameterised by the first-order logic are defined in the same way as in Definitions 5.5.1 and 5.1.2, except for the replacement of Rule (ii) with the following rule and the addition of the semantics for the following formulae:

- The set of formulae in $\mathbb{L}(\Upsilon, FO)$ is closed under
  - first-order logical connectives $\neg, \wedge, \vee, \Rightarrow, \Leftrightarrow$ (i.e., $\neg\psi, \psi_1 \wedge \psi_2, \psi_1 \vee \psi_2, \psi_1 \Rightarrow \psi_2, \psi_1 \Leftrightarrow \psi_2$) where
    * $fr(\neg(\varphi)) = fr(\varphi)$, and
    * $fr(\varphi_1\ con\ \varphi_2) = fr(\varphi_1) \cup fr(\varphi_2)$ for $con \in \{\wedge, \vee, \Rightarrow, \Leftrightarrow\}$,
  - first-order quantifiers $\exists$ and $\forall$ (i.e., $\exists\overline{x}.\varphi$ and $\forall\overline{x}.\varphi$) where
    * $fr(\exists\overline{x}.\varphi) = fr(\varphi) - \{\overline{x}\}$, and
    * $fr(\forall\overline{x}.\varphi) = fr(\varphi) - \{\overline{x}\}$.
- The semantics of these formulae is defined by
  - $[\![\neg\psi]\!]_{S,\zeta} = \begin{cases} true & \text{if } [\![\psi]\!]_{S,\zeta} = false \\ false & \text{otherwise} \end{cases}$
  - $[\![\psi_1 \wedge \psi_2]\!]_{S,\zeta} = \begin{cases} true & \text{if } [\![\psi_1]\!]_{S,\zeta} = true \text{ and } [\![\psi_2]\!]_{S,\zeta} = true \\ false & \text{otherwise} \end{cases}$

- $[\![\psi_1 \vee \psi_2]\!]_{S,\varsigma} = \begin{cases} true & \text{if } [\![\psi_1]\!]_{S,\varsigma} = true \text{ or } [\![\psi_2]\!]_{S,\varsigma} = true \\ false & \text{otherwise} \end{cases}$
- $[\![\psi_1 \Rightarrow \psi_2]\!]_{S,\varsigma} = \begin{cases} true & \text{if } [\![\psi_1]\!]_{S,\varsigma} = false \text{ or } [\![\psi_2]\!]_{S,\varsigma} = true \\ false & \text{otherwise} \end{cases}$
- $[\![\psi_1 \Leftrightarrow \psi_2]\!]_{S,\varsigma} = \begin{cases} true & \text{if both } [\![\psi_1]\!]_{S,\varsigma} \text{ and } [\![\psi_2]\!]_{S,\varsigma} \text{ are } true\text{, or} \\ & \quad \text{both} [\![\psi_1]\!]_{S,\varsigma} \text{ and } [\![\psi_2]\!]_{S,\varsigma} \text{ are } false \\ false & \text{otherwise} \end{cases}$
- $[\![\exists x.\varphi]\!]_{S,\varsigma} = \begin{cases} true & \text{if } [\![\varphi]\!]_{S,\varsigma[x\mapsto a]} = true \text{ for at least one } a \in B_{db} \\ false & \text{otherwise} \end{cases}$
- $[\![\forall x.\varphi]\!]_{S,\varsigma} = \begin{cases} true & \text{if } [\![\varphi]\!]_{S,\varsigma[x\mapsto a]} = true \text{ for every } a \in B_{db} \\ false & \text{otherwise} \end{cases}$

The syntax and semantics of $\mathbb{L}_M^{FO}$ are defined on top of $\mathbb{L}(\Upsilon, FO)$ as in Definitions 5.2.1 and 5.2.2.

5.5.1 Translation to FO Logic

Before presenting the completeness proofs for the proof system of $\mathbb{L}_M^{FO}$, we need to show how second-order formulae in $\mathbb{L}_M^{FO}$ can be translated into a many-sorted first-order logic. It is well-known that a many-sorted first-order logic can again be reduced to a one-sorted first-order logic [67]. Therefore, although second-order formulae are used in $\mathbb{L}_M^{FO}$, they do not increase the expressive power of the logic. Instead, we add them only for the sake of convenience and conciseness.

The following approach for translating $\mathbb{L}_M^{FO}$ to a many-sorted first-order logic is quite standard. Let us consider a many-sorted structure in which we have

- an *individual sort* (with variables $x_1, x_2, \ldots$), and
- many *predicate sorts* such that for each $n \in \mathbb{N}$, there is a n-ary predicate sort (with variables $X_1^n, X_2^n, \ldots$).

An universe of the n-ary predicate sort is a set of n-ary relations over the universe of the individual sort. The terms in $\mathbb{L}_M^{FO}$ are all of the individual sort, except for variables $X_1^n, X_2^n, \ldots$ that are of the n-ary predicate sort.

The equality and inequality are used under a fixed interpretation and only between terms of the individual sort. Furthermore, For each $n \in \mathbb{N}$, there is a membership predicate $\in^n$ which has the $n+1$ arguments of sort: the n-ary predicate sort and n individual sorts such that

$$\in^n (X^n, x_1, ..., x_n)$$

with the following semantics:

$$[\![\in^n (X^n, x_1, ..., x_n)]\!]_{S,\zeta} = \begin{cases} true & \text{if } ([\![x_1]\!]_{S,\zeta}, ..., [\![x_n]\!]_{S,\zeta}) \in [\![X^n]\!]_{S,\zeta}), \\ false & \text{otherwise.} \end{cases}$$

Based on the use of predicate sorts and membership predicates, now we can translate those extended formulae with the second-order syntax in $\mathbb{L}_M^{FO}$ to formulae in a many-sorted first-order logic, where X^n is of the n-arity predicate sort, and x_i is of the individual sort:

- $\exists X.\varphi$ and $\forall X.\varphi$ can be translated to
  - $\exists X^3.\varphi$ and $\forall X^3.\varphi$ if X is bound to an update set, or
  - $\exists X^4.\varphi$ and $\forall X^4.\varphi$ if X is bound to an update multiset.
- upd(r, Δ) can be translated to upd(r, X^3_Δ)
- upm$(r, \ddot{\Delta})$ can be translated to upm$(r, X^4_{\ddot{\Delta}})$
- $\Delta(f, \overline{x}, y)$ can be translated to $\in^3_\Delta (X^3, x_1, ..., x_3)$
- $\ddot{\Delta}(f, \overline{x}, y, z)$ can be translated to $\in^4_{\ddot{\Delta}} (X^4, x_1, ..., x_4)$

Without loss of generality, in the completeness proofs discussed in the following subsections, we will treat these second-order formulae as being many-sorted first-order formulae in disguise.

Furthermore, in accordance with the approach used in [47], we associate a rank $|\varphi| \in \mathbb{N}$ with each formula φ and a rank $|r| \in \mathbb{N}$ with each DB-ASM rule r, such that

- $|t_1 = t_2| = 0$
- $|P_1(t_1, ..., t_n)| = |P_2(t_1, ..., t_n)| = 0$

- $|\Delta(f, \overline{x}, y)| = 0$
- $|\ddot{\Delta}(f, \overline{x}, y, z)| = 0$
- $|\neg\varphi| = |\varphi| + 1$
- $|\varphi_1 \; con \; \varphi_2| = \max(|\varphi_1, \varphi_2|) + 1$ for $con \in \{\wedge, \vee, \Rightarrow\}$
- $|\forall\overline{x}.\varphi| = |\exists\overline{x}.\varphi| = |\varphi| + 1$
- $|\exists X.\varphi| = |\forall X.\varphi| = |\varphi| + 1$
- $|\mathrm{upd}(r, \Delta)| = |\varphi| + 1$, if $\varphi \Leftrightarrow \mathrm{upd}(r, \Delta)$ is an instance of Axioms **U1-U7**
- $|\mathrm{upm}(r, \ddot{\Delta})| = |\varphi| + 1$, if $\varphi \Leftrightarrow \mathrm{upm}(r, \ddot{\Delta})$ is an instance of Axioms **UM1-UM7**
- $|[\Delta]\varphi| = |r| + |\varphi| + 1$, if Δ is generated by rule r
- $|r| = \max(|\mathrm{con}(r, \Delta)|, |\mathrm{upd}(r, \Delta)|, |\mathrm{upm}(r, \ddot{\Delta})|, |\mathrm{inv}(\Delta, f, \overline{x})|) + 1$

5.5.2 Henkin Construction

In this subsection we present the completeness proof for the proof system of $\mathbb{L}_M^{FO}$ based on the Henkin model construction.

Let $\mathcal{X}_H = \{x_{h_1}, x_{h_2}, ...\}$ be a countable set of fresh variables which is disjoint from the signature Υ_M of a DB-ASM M. Variables in $\mathcal{X}_H$ serve as Henkin constants as in the classical Henkin style completeness proof. By adjoining $\mathcal{X}_H$ to Υ_M, we expand the signature Υ_M with Henkin constants. That is, $\Upsilon_{MH} = \Upsilon_M \cup \mathcal{X}_H$.

Definition 5.5.2. A set Ψ of formulae is *satisfiable* iff there exists a state S and a variable assignment ζ such that $[\![\varphi]\!]_{S,\zeta} = \mathit{true}$ for each $\varphi \in \Psi$.

Definition 5.5.3. A set Ψ of formulae is *inconsistent* iff there exists a finite subset $\Psi' \subseteq \Psi$ such that $\bigwedge_{\varphi \in \Psi'} \varphi \Rightarrow \bot$ is derivable; otherwise, Ψ is *consistent.*

Definition 5.5.4. A set Ψ of formulae is *maximally consistent* iff the following two statements are true:

- Ψ is consistent.
- If $\Psi \cup \{\varphi\}$ is consistent, then $\varphi \in \Psi$.

The set Ψ of formulae over Υ_M can be extended to the set Ψ' of formulae over Υ_{MH} by using the following rule:

* For every formula $\exists x.\varphi \in \Psi$, there is a Henkin constant $x_h \in \mathcal{X}_H$ such that $\varphi[x_h/x] \in \Psi'$.

The above extension preserves the consistency of the set Ψ of formulae. Put it in other words, if Ψ is consistent, then Ψ' must also be consistent.

Definition 5.5.5. The set Ψ' of formulae over Υ_{MH} is said to *contain witnesses* iff for each formula $\exists x.\varphi \in \Psi'$ there exists $\varphi[x_h/x] \in \Psi'$.

Furthermore, any set Ψ' of consistent formulae over Υ_{MH} can be further expended to a set Ψ'' of maximally consistent formulae over Υ_{MH} such that $\Psi' \supseteq \Psi''$. Inductively substituting quantified variables with Henkin constants may lead to a transformation from formulae into sentences.

Lemma 5.5.1. *For each maximal consistent set Ψ of formulae, it satisfies the following statements.*

(1) For each formula φ, either $\varphi \in \Psi$ or $\neg\varphi \in \Psi$.

(2) If $\varphi \in \Psi$ and $\varphi \Rightarrow \psi$ is derivable, then $\psi \in \Psi$.

Proof. The proof is straightforward.

- For Statement (1), assume that there is a formula φ' such that $\varphi' \notin \Psi$ and $\neg\varphi' \notin \Psi$. By Definition 5.5.4 and $\varphi' \notin \Psi$, we know that $\Psi \cup \{\varphi'\}$ is not consistent. Therefore, $\Psi \cup \{\neg\varphi'\}$ must be consistent, which means that $\neg\varphi' \in \Psi$. There is a contradiction.

- For Statement (2), we assume that $\varphi \in \Psi$ and $\varphi \Rightarrow \psi$ is derivable, but $\psi \notin \Psi$. Since $\psi \notin \Psi$, then $\neg\psi \in \Psi$. It mean that $\Psi \cup \{\neg\psi\}$ is consistent. By the condition that $\varphi \in \Psi$, we have $\{\varphi, \neg\psi\} \subseteq \Psi$. However, because $\varphi \Rightarrow \psi$ holds, $\varphi \wedge \neg\psi \Rightarrow \bot$ is derivable. This contradicts with the condition that Ψ is a maximal consistent set of formulae.

□

Let Ψ be a set of formulae and r be a DB-ASM rule yielding an update set Δ. Then we define Ψ_r^Δ to be a set of formulae such that

$$\Psi_r^\Delta = \{\varphi | [\Delta]\varphi \in \Psi\}.$$

It means that, if Ψ is the set of formulae which are true in a state S and the update set Δ generated by the rule r over S is consistent, then Ψ_r^Δ is the set of formulae which are true in a state $S + \Delta$, i.e., a state after firing the update set Δ over the state S.

Lemma 5.5.2. *Let Ψ be a maximal consistent set of formulae that contains witnesses. If $con(r, \Delta) \in \Psi$, then Ψ_r^Δ is a maximal consistent set of formulae, containing witnesses.*

Proof. The proof will be proceeded in three steps.

(1) The first step is to prove that Ψ_r^Δ is consistent. Assume that Ψ_r^Δ is inconsistent and there is a finite subset $\Psi' \subseteq \Psi_r^\Delta$ of formulae such that $\bigwedge_{\varphi \in \Psi'} \varphi \Rightarrow \bot$ is derivable. By Axiom **M1**, we get $[\Delta](\bigwedge_{\varphi \in \Psi'} \varphi \Rightarrow \bot) \Rightarrow ([\Delta] \bigwedge_{\varphi \in \Psi'} \varphi \Rightarrow [\Delta]\bot)$. By Rule **M2**, $(\bigwedge_{\varphi \in \Psi'} \varphi \Rightarrow \bot) \Rightarrow [\Delta](\bigwedge_{\varphi \in \Psi'} \varphi \Rightarrow \bot)$ is derivable. Hence, by Rule **M3**, we have $[\Delta] \bigwedge_{\varphi \in \Psi'} \varphi \Rightarrow [\Delta]\bot$. By Property (8) in Lemma 5.4.2, $\bigwedge_{\varphi \in \Psi'} [\Delta]\varphi \Rightarrow [\Delta] \bigwedge_{\varphi \in \Psi'} \varphi \Rightarrow [\Delta]\bot$. According to the definition of Ψ_r^Δ, the formula $[\Delta]\varphi$ is in Ψ for each $\varphi \in \Psi'$, and so $\bigwedge_{\varphi \in \Psi'} [\Delta]\varphi \Rightarrow \bot$ should not be derivable. However, by an instance of Axiom **M8** (i.e., $\mathrm{con}(r, \Delta) \wedge [\Delta]\bot \Rightarrow \bot$) and the fact that $\mathrm{con}(r, \Delta) \in \Psi$ is the assumption of this lemma, we can derive $\bigwedge_{\varphi \in \Psi'} [\Delta]\varphi \Rightarrow \bot$. There is a contradiction. Thus, Ψ_r^Δ is consistent.

(2) Secondly, we need to prove that Ψ_r^Δ is maximal consistent. That is, if $\Psi_r^\Delta \cup \{\varphi\}$ is consistent, then we need to show that $\varphi \in \Psi_r^\Delta$. More precisely, by the definition of Ψ_r^Δ, we have to show that $[\Delta]\varphi \in \Psi$. Since Ψ is maximal consistent, by Statement (1) of Lemma 5.5.1, we know that either $[\Delta]\varphi \in \Psi$ or $\neg[\Delta]\varphi \in \Psi$.

- Assume that $[\Delta]\varphi \in \Psi$. Then the proof for $\varphi \in \Psi_r^\Delta$ finishes because that is the result we desire.
- Assume that $\neg[\Delta]\varphi \in \Psi$. By the instance $\neg[\Delta]\varphi \Rightarrow [\Delta]\neg\varphi$ of Axiom **M5** and Statement (2) of Lemma 5.5.1 that Ψ is deductively closed, we can get $[\Delta]\neg\varphi \in \Psi$. By the definition of Ψ_r^Δ, it follows that $\neg\varphi \in \Psi_r^\Delta$. However, this implies that $\Psi_r^\Delta \cup \{\neg\varphi\}$ is consistent, which contradicts with the assumption that $\Psi_r^\Delta \cup \{\varphi\}$ is consistent. Thus, this case is not possible.

Based on the above results, we conclude that Ψ_r^Δ is maximal consistent.

(3) Finally, we need to prove that Ψ_r^Δ contains witnesses. Assume that $\exists x.\varphi \in \Psi_r^\Delta$ for $x \notin fr(r)$. Then by the definition of Ψ_r^Δ, we get $[\Delta]\exists x.\varphi \in \Psi$. By Property (7) in Lemma 5.4.2 that shows $[\Delta]\exists x.\varphi \Rightarrow \exists x.[\Delta]\varphi$ for $x \notin fr(r)$, and Statement (2) of Lemma 5.5.1 that Ψ is deductively closed, it follows that $\exists x.[\Delta]\varphi \in \Psi$. Because Ψ contains witnesses, there exists a Henkin constant x_h such that $[\Delta]\varphi[x_h/x] \in \Psi$. By the definition of Ψ_r^Δ, it implies that $\varphi[x_h/x] \in \Psi_r^\Delta$. The proof completes.

□

In terms of a maximal consistent set Ψ of formulae containing witnesses, we adopt the approach presented in [47] to construct a state S_Ψ. First of all, we define an *equivalence relation* $\sim_\Psi$ on the set of variables such that

$$x \sim_\Psi y \equiv (x = y) \in \Psi.$$

As a convention, we use $\lceil x \rceil_\Psi$ to denote the equivalence class of a variable x and $\lceil \sim_\Psi \rceil$ to denote the set of all possible equivalence classes with respect to $\sim_\Psi$ in a state, respectively. Let S_Ψ be a state which has the set $\lceil \sim_\Psi \rceil$ as its universe. For two equivalence classes $\lceil x \rceil_\Psi$ and $\lceil y \rceil_\Psi$ in $\lceil \sim_\Psi \rceil$, we define that

$$f^{S_\Psi}(\lceil x \rceil_\Psi) = \lceil y \rceil_\Psi \equiv \text{there exist } x \in \lceil x \rceil_\Psi \text{ and } y \in \lceil y \rceil_\Psi \text{ such that } (f(x) = y) \in \Psi.$$

The function f^{S_Ψ} is well-defined and total the same as discussed in [47]. That is, by the derivable property of functions, i.e., $(f(x_1) = y_1 \wedge f(x_2) = y_2 \wedge x_1 = x_2) \Rightarrow y_1 = y_2$, and Statement (2) of Lemma 5.5.1 that Ψ is deductively closed, we have,

$$(f^{S_\Psi}(\lceil x_1 \rceil_\Psi) = \lceil y_1 \rceil_\Psi \wedge f^{S_\Psi}(\lceil x_2 \rceil_\Psi) = \lceil y_2 \rceil_\Psi \wedge \lceil x_1 \rceil_\Psi = \lceil x_2 \rceil_\Psi) \Rightarrow \lceil y_1 \rceil_\Psi = \lceil y_2 \rceil_\Psi.$$

Furthermore, as Ψ contains witness, it means that, for each variable x, we have $f(x) = x_h$ in Ψ with a Henkin constant x_h. In this way, the function f^{S_Ψ} can be treated as being total.

Now we need to deal with terms. Since the terms of the logic $\mathbb{L}_M^{FO}$ for DB-ASMs are the same as the terms of the logic $\mathbb{L}(\Upsilon, FO)$ of meta-finite states with the same signature, it means that ρ-terms have to be handled in the construction of a state S_Ψ. From the syntax and semantics of ρ-terms defined in Subsection 5.1.1, we know that the use of ρ-terms may lead to nested constructions between formulae and terms. Therefore, we need to start with the construction of a state S_Ψ where the ρ-ranks of all formulae in Ψ are 0 (i.e., no ρ-terms in any formula of Ψ) and prove Lemma 5.5.3 based on the construction for terms that do not contain any ρ-terms. After that, we extend the results of Lemma 5.5.3 to those in Lemma 5.5.4, in which the state S_{Ψ_ρ} is constructed from terms that may have ρ-ranks at most $n \in \mathbb{N}$ by using inductive arguments for constructing states level-by-level with respect to ρ-ranks.

Let us begin with pure terms, i.e., terms that do not contain ρ-terms. We use $\widetilde{\zeta}$ for a variable assignment which assigns an equivalence class $\lceil x \rceil_\Psi$ to a variable x, $[\![t]\!]_{\Psi,\widetilde{\zeta}}$ and $[\![\varphi]\!]_{\Psi,\widetilde{\zeta}}$ for the interpretations of a pure term t and a formula φ in state S_Ψ under $\widetilde{\zeta}$, respectively. For a pure term t, the following property holds.

$$[\![t]\!]_{\Psi,\widetilde{\zeta}} = \lceil x \rceil_\Psi \equiv (t = x) \in \Psi$$

Since for each pure term t, the formula $t = x$ for a variable x may belong to Ψ, the set of equivalence classes of terms is isomorphic to the set of equivalence classes of variables.

By treating natural numbers in the state S_Ψ as terms, the interpretation of functions $\leq$ and $+$ over natural numbers can be fixed, such that

- $([\![t_1 \leq t_2]\!]_{\Psi,\widetilde{\zeta}} \wedge [\![t_1]\!]_{\Psi,\widetilde{\zeta}} = \lceil x_1 \rceil_\Psi \wedge [\![t_2]\!]_{\Psi,\widetilde{\zeta}} = \lceil x_2 \rceil_\Psi) \Rightarrow \lceil x_1 \rceil_\Psi \leq \lceil x_2 \rceil_\Psi$, and
- $([\![t_1 + t_2 = t_3]\!]_{\Psi,\widetilde{\zeta}} \wedge [\![t_1]\!]_{\Psi,\widetilde{\zeta}} = \lceil x_1 \rceil_\Psi \wedge [\![t_2]\!]_{\Psi,\widetilde{\zeta}} = \lceil x_2 \rceil_\Psi \wedge [\![t_3]\!]_{\Psi,\widetilde{\zeta}} = \lceil x_3 \rceil_\Psi) \Rightarrow \lceil x_1 \rceil_\Psi + \lceil x_2 \rceil_\Psi = \lceil x_3 \rceil_\Psi$.

Based on the equivalence classes of variables, the property of terms and the fixed interpretation of functions $\leq$ and $+$ as discussed before, all the possible update sets and multisets in the state S_Ψ are determined. Thus, by the finiteness of update sets and multisets, the interpretation of $\in^3_\Delta$ and $\in^4_{\ddot{\Delta}}$ is fixed in the state S_Ψ as well.

- The interpretation of the formula $\Delta(f, \bar{t}, t_0)$ in the state S_Ψ under $\widetilde{\zeta}$ has the following property:
  - $(\Delta(f, [\![\bar{t}]\!]_{\Psi,\widetilde{\zeta}}, [\![t_0]\!]_{\Psi,\widetilde{\zeta}}) \wedge [\![\bar{t}]\!]_{\Psi,\widetilde{\zeta}} = \lceil \bar{x} \rceil_\Psi \wedge [\![t_0]\!]_{\Psi,\widetilde{\zeta}} = \lceil x_0 \rceil_\Psi) \Rightarrow \Delta(f, \lceil \bar{x} \rceil_\Psi, \lceil x_0 \rceil_\Psi)$, and
  - $\Delta(f, \lceil \bar{x} \rceil_\Psi, \lceil x_0 \rceil_\Psi) \Rightarrow \in^3_\Delta (\Delta, f, \lceil \bar{x} \rceil_\Psi, \lceil x_0 \rceil_\Psi)$
- Similarly, the interpretation of the formula $\ddot{\Delta}(f, \bar{t}, t_0, t')$ in the state S_Ψ under $\widetilde{\zeta}$ has the following property:
  - $(\ddot{\Delta}(f, [\![\bar{t}]\!]_{\Psi,\widetilde{\zeta}}, [\![t_0]\!]_{\Psi,\widetilde{\zeta}}, [\![t']\!]_{\Psi,\widetilde{\zeta}}) \wedge [\![\bar{t}]\!]_{\Psi,\widetilde{\zeta}} = \lceil \bar{x} \rceil_\Psi \wedge [\![t_0]\!]_{\Psi,\widetilde{\zeta}} = \lceil x_0 \rceil_\Psi \wedge [\![t']\!]_{\Psi,\widetilde{\zeta}} = \lceil x' \rceil_\Psi) \Rightarrow \ddot{\Delta}(f, \lceil \bar{x} \rceil_\Psi, \lceil x_0 \rceil_\Psi, \lceil x' \rceil_\Psi)$, and

– $\ddot{\Delta}(f, \lceil\overline{x}\rceil_\Psi, \lceil x_0\rceil_\Psi, \lceil x'\rceil_\Psi) \Rightarrow \in^4_{\ddot{\Delta}} (\ddot{\Delta}, f, \lceil\overline{x}\rceil_\Psi, \lceil x_0\rceil_\Psi, \lceil x'\rceil_\Psi)$

Lemma 5.5.3. *Let φ be a formula. For any maximal consistent set Ψ of formulae which contains witnesses but does not contain any ρ-terms, the following two statements are true:*

(a). *If $\varphi \in \Psi$, then $[\![\varphi]\!]_{\Psi,\widetilde{\zeta}} = true$.*

(b). *If $\neg\varphi \in \Psi$, then $[\![\varphi]\!]_{\Psi,\widetilde{\zeta}} = false$.*

Proof. In a similar way to the approach used in [47], the proof is based on the induction on the rank of a formula.

◦ **The case of formula $[\Delta]\varphi$**

Assume that Statements (a) and (b) are true for all formulae that have ranks less than $|[\Delta]\varphi|$. Then we first need to prove the following statement for a DB-ASM rule r.

§. If $\text{con}(r, \Delta) \in \Psi$, then Δ is consistent and $S_\Psi + \Delta = S_{\Psi^\Delta_r}$.

Statement § says that if the formula $\text{con}(r, \Delta)$ is in Ψ, then Δ is a consistent update set yielded by r in S_Ψ and the state $S_{\Psi^\Delta_r}$ associated with the set Ψ^Δ_r of formulae is equal to a state obtained by firing the update set Δ over S_Ψ.

As $\text{con}(r, \Delta) \in \Psi$ and $|\text{con}(r, \Delta)| < |[\Delta]\varphi|$, by the induction hypothesis (a) for $\text{con}(r, \Delta) \in \Psi$, we get $[\![\text{con}(r, \Delta)]\!]_{\Psi,\widetilde{\zeta}} = true$. It means that by definition that Δ is a consistent update set yielded by applying rule r over S_Ψ under $\widetilde{\zeta}$. Then, by Axiom **M7** for static and pure formula $x = y$, $(x = y) \Rightarrow [\Delta](x = y)$ can be derived. By Statement (2) of Lemma 5.5.1 that Ψ is deductively closed, we know that, if $x = y \in \Psi$, then $[\Delta](x = y) \in \Psi$. By the definition of Ψ^Δ_r, it follows that $(x = y) \in \Psi^\Delta_r$. Therefore, the equivalent relation $\sim_\Psi$ is the same as the equivalent relation $\sim_{\Psi^\Delta_r}$ and the universe of the structure S_Ψ is the same as the universe of the structure $S_{\Psi^\Delta_r}$. That is, $\lceil\sim_\Psi\rceil = \lceil\sim_{\Psi^\Delta_r}\rceil$.

Let x be an arbitrarily chosen variable. Now we have a look at the formula $\text{inv}(\Delta, f, \overline{x})$. By Statement (1) of Lemma 5.5.1, we know that either $\text{inv}(\Delta, f, \overline{x})$ or $\neg\text{inv}(\Delta, f, \overline{x})$ belongs to Ψ.

- Assume that $\text{inv}(\Delta, f, \overline{x}) \in \Psi$. Since $|\text{inv}(\Delta, f, \overline{x})| < |[\Delta]\varphi|$, by the induction hypothesis (a) for $\text{inv}(\Delta, f, \overline{x}) \in \Psi$, we can get $[\![\text{inv}(\Delta, f, \overline{x})]\!]_{\Psi,\tilde{\zeta}} = true$. It means that there is no update for f at the argument $\lceil \overline{x} \rceil_\Psi$ in the update set Δ. So, $f^{S_\Psi+\Delta}(\lceil \overline{x} \rceil_\Psi) = \lceil y \rceil_\Psi$ holds in the state after applying Δ over S_Ψ, where y is a variable such that $(f(\overline{x}) = y) \in \Psi$. By Axiom **A1** and Statement (2) of Lemma 5.5.1, we obtain $[\Delta]f(\overline{x}) = y \in \Psi$. By the definition of Ψ_r^Δ, $(f(\overline{x}) = y) \in \Psi_r^\Delta$. Therefore, $f^{S_{\Psi_r^\Delta}}(\lceil \overline{x} \rceil_\Psi) = \lceil y \rceil_\Psi$ and so f has the same value at $\lceil \overline{x} \rceil_\Psi$ in both the structure $S_\Psi + \Delta$ and the structure $S_{\Psi_r^\Delta}$.
- Assume that $\neg\text{inv}(\Delta, f, \overline{x}) \in \Psi$. By the definition of $\text{inv}(\Delta, f, \overline{x})$, it means that $\exists y \Delta(f, \overline{x}, y) \in \Psi$. As Ψ contains witnesses, there exists a Henkin constant x_h such that $\Delta(f, \overline{x}, x_h) \in \Psi$. Since $|\Delta(f, \overline{x}, x_h)| < |[\Delta]\varphi|$, by the induction hypothesis (a) for $\Delta(f, \overline{x}, x_h) \in \Psi$, we know that $[\![\Delta(f, \overline{x}, x_h)]\!]_{\Psi,\tilde{\zeta}} = true$. Therefore, $((f, \lceil \overline{x} \rceil_\Psi), \lceil x_h \rceil_\Psi) \in \Delta$ and so $f^{S_\Psi+\Delta}(\lceil \overline{x} \rceil_\Psi) = \lceil x_h \rceil_\Psi$. According to Axiom **A2**, we get $[\Delta]f(\overline{x}) = x_h \in \Psi$. Then, by the definition of Ψ_r^Δ, we know that $f(\overline{x}) = x_h \in \Psi_r^\Delta$. It means that $f^{S_{\Psi_r^\Delta}}(\lceil \overline{x} \rceil_\Psi) = \lceil x_h \rceil_\Psi$ and the function f also has the same value $\lceil \overline{x} \rceil_\Psi$ in both the structures $S_\Psi + \Delta$ and the structure $S_{\Psi_r^\Delta}$.

So far, we have completed the proof for Statement §. In the following, we will continue to prove the induction hypotheses (a) and (b) for the formulae $[\Delta]\varphi$, where Δ is an update set yielded by r.

- Assume that $[\Delta]\varphi \in \Psi$. By Statement (1) of Lemma 5.5.1, we know that either $\text{con}(r, \Delta) \in \Psi$ or $\neg\text{con}(r, \Delta) \in \Psi$.
  - Assume that $\text{con}(r, \Delta) \in \Psi$. By Lemma 5.5.2, Ψ_r^Δ is maximal consistent, containing witnesses. From the definition of Ψ_r^Δ, it follows that $\varphi \in \Psi_r^\Delta$. Since $|\varphi| < |[\Delta]\varphi|$, by the induction hypothesis (a) for $\varphi \in \Psi_r^\Delta$, it follows that $[\![\varphi]\!]_{\Psi_r^\Delta,\tilde{\zeta}} = true$. Thus, by Statement §, we obtain $[\![[\Delta]\varphi]\!]_{\Psi,\tilde{\zeta}} = true$.
  - Assume that $\neg\text{con}(r, \Delta) \in \Psi$. Since $|\text{con}(r, \Delta)| < |[\Delta]\varphi|$, by the induction hypothesis (b) for $\text{con}(r, \Delta)$, it follows that $[\![\text{con}(r, \Delta)]\!]_{\Psi,\tilde{\zeta}} = false$. Therefore, we obtain $[\![[\Delta]\varphi]\!]_{\Psi,\tilde{\zeta}} = true$.

- Assume that $\neg[\Delta]\varphi \in \Psi$. Since Ψ is deductively closed by Statement (2) of Lemma 5.5.1 and Ψ is consistent, by Axioms **M4** and **M5**, we have $[\Delta]\neg\varphi \in \Psi$ and $\text{con}(r, \Delta) \in \Psi$. Because $\text{con}(r, \Delta) \in \Psi$, by Lemma 5.5.2, it means that Ψ_r^Δ is maximal consistent containing witnesses. By the definition of Ψ_r^Δ, we have

$\neg\varphi \in \Psi_r^{\Delta}$. By the induction hypothesis (b), it follows that $[\![\varphi]\!]_{\Psi_r^{\Delta},\tilde{\zeta}} = false$. By Statement § again, we obtain $[\![[\Delta]\varphi]\!]_{\Psi,\tilde{\zeta}} = false$.

∘ **The case of formulae upd(r, Δ) and upm$(r, \ddot{\Delta})$**

- upd(r, Δ):

  As discussed in Subsection 5.5.1, second-order variables for Δ can be translated to variables of the 3-arity predicate sort in a many-sorted first-order logic. Because Ψ contains witnesses, there exists a variable X_{Δ}^3 such that the equation $X_{\Delta}^3 = \Delta \in \Psi$. Then for each formula upd(r, X_{Δ}^3), there exists a formula φ such that upd$(r, X_{\Delta}^3) \Leftrightarrow \varphi$ is an instance of Axioms **U1-U7**. By the definition of the ranks of formulae, we know that

  $$|\varphi| < |\text{upd}(r, X_{\Delta}^3)|.$$

  - Assume that upd$(r, X_{\Delta}^3) \in \Psi$. Since Ψ is deductively closed according to Statement (2) of Lemma 5.5.1, the formula φ is also in Ψ. By the induction hypothesis (a) for $\varphi \in \Psi$, it follows that $[\![\varphi]\!]_{\Psi,\tilde{\zeta}} = true$. As upd$(r, X_{\Delta}^3) \Leftrightarrow \varphi$ is valid in any structure, so $[\![\text{upd}(r, X_{\Delta}^3)]\!]_{\Psi,\tilde{\zeta}} = true$.
  - Assume that $\neg$upd$(r, X_{\Delta}^3) \in \Psi$. Similarly, since Ψ is deductively closed according to Statement (2) of Lemma 5.5.1, the formula $\neg\varphi$ is also in Ψ. By the induction hypothesis (b) for $\neg\varphi \in \Psi$, it follows that $[\![\varphi]\!]_{\Psi,\tilde{\zeta}} = false$. As upd$(r, X_{\Delta}^3) \Leftrightarrow \varphi$ is valid in any structure, so $[\![\text{upd}(r, X_{\Delta}^3)]\!]_{\Psi,\tilde{\zeta}} = false$.

- upm$(r, \ddot{\Delta})$:

  The proof for upm$(r, \ddot{\Delta})$ is similar to the proof for upd(r, Δ) as above.

  As discussed in Subsection 5.5.1, second-order variables for $\ddot{\Delta}$ can be translated to variables of the 4-arity predicate sort in a many-sorted first-order logic. Because Ψ contains witnesses, there exists a variable $X_{\ddot{\Delta}}^4$ such that the equation $X_{\ddot{\Delta}}^4 = \ddot{\Delta} \in \Psi$. Then for each formula upm$(r, X_{\ddot{\Delta}}^4)$, there exists a formula φ' such that upm$(r, X_{\ddot{\Delta}}^4) \Leftrightarrow \varphi'$ is an instance of Axioms **UM1-UM7**. By the definition of the ranks of formulae, we know that

  $$|\varphi'| < |\text{upm}(r, X_{\ddot{\Delta}}^4)|.$$

  - Assume that upm$(r, X_{\ddot{\Delta}}^4) \in \Psi$. Since Ψ is deductively closed according to Statement (2) of Lemma 5.5.1, the formula φ' is also in Ψ. By the induction hypothesis (a) for $\varphi' \in \Psi$, it follows that $[\![\varphi']\!]_{\Psi,\tilde{\zeta}} = true$. As $\mathrm{upm}(r, X^4_\Delta) \Leftrightarrow \varphi'$ is valid in any structure, so $[\![\mathrm{upm}(r, X^4_\Delta)]\!]_{\Psi,\tilde{\zeta}} = true$.

- Assume that $\neg\mathrm{upm}(r, X^4_\Delta) \in \Psi$. Similarly, since Ψ is deductively closed according to Statement (2) of Lemma 5.5.1, the formula $\neg\varphi'$ is also in Ψ. By the induction hypothesis (b) for $\neg\varphi' \in \Psi$, it follows that $[\![\varphi']\!]_{\Psi,\tilde{\zeta}} = false$. As $\mathrm{upm}(r, X^4_\Delta) \Leftrightarrow \varphi'$ is valid in any structure, so $[\![\mathrm{upm}(r, X^4_\Delta)]\!]_{\Psi,\tilde{\zeta}} = false$.

□

In the previous discussion, the property $[\![t]\!]_{\Psi,\tilde{\zeta}} = \lceil x \rceil_\Psi \equiv (t = x) \in \Psi$ holds for a pure term t. Now we need to extend the results of Lemma 5.5.3 to the more general case, i.e., a state that may have ρ-terms. The approach we use is to first treat a ρ-term $\rho_{\overline{x}}(t|\varphi_1(\overline{x}, \overline{y}))$ as a term $t_1(\overline{y})$ with free variables $\overline{y}$. That is, i.e.,

$$t_1(\overline{y}) \equiv \rho_{\overline{x}}(t|\varphi_1(\overline{x}, \overline{y})).$$

Then for each ρ-term $\rho_{\overline{x}}(t|\varphi_1(\overline{x}, \overline{y}))$, we have the following formula

$$t_1(\overline{y}) = z \wedge \exists\overline{x}.\varphi_1(\overline{x}, \overline{y}).$$

As ρ-terms may be nested inductively, the above formula needs to be further extended if t or $\varphi_1(\overline{x}, \overline{y})$ in the ρ-term $\rho_{\overline{x}}(t|\varphi_1(\overline{x}, \overline{y}))$ has other ρ-terms. Assume that there is another ρ-term $\rho'_{\overline{x}'}(t'|\varphi_2(\overline{x}', \overline{y}'))$ included in t or $\varphi_1(\overline{x}, \overline{y})$. Then by using the shortcut $t_2(\overline{y}')$ for this ρ-term, such that

$$t_2(\overline{y}') \equiv \rho'_{\overline{x}'}(t'|\varphi_2(\overline{x}', \overline{y}')),$$

we can extend the above formula to the following:

$$t_1(\overline{y}) = z \wedge \exists\overline{x}.\varphi'_1(\overline{x}, \overline{y}, z') \wedge t_2(\overline{y}') = z' \wedge \exists\overline{x}'.\varphi_2(\overline{x}', \overline{y}').$$

with

$$\varphi'_1(\overline{x}, \overline{y}, z') \equiv \varphi_1(\overline{x}, \overline{y})[z'/\rho'_{\overline{x}'}(t'|\varphi_2(\overline{x}', \overline{y}'))].$$

In doing so, we can flatten the nesting constructions of each ρ-term into a set of formulae over a state. In other words, for every set Ψ of formulae which contains witnesses and ρ-terms, we can extend it to a set Ψ_ρ of formulae which contains witnesses and formulae in ρ-terms by using the above approach, without loss of generality. Furthermore, this extension preserves the consistency of the set Ψ of formulae as well. That is, if Ψ is consistent, then Ψ_ρ must also be consistent.

Now based on the set Ψ_ρ of formulae which contains witnesses and is maximal consistent, we can define an equivalence relation $\sim_{\Psi_\rho}$ on the set of variables such that

$$x \sim_{\Psi_\rho} y \equiv (x = y) \in \Psi_\rho.$$

Similarly, the equivalence class of a variable x is denoted as $\lceil x \rceil_{\Psi_\rho}$ and the set of all possible equivalence classes with respect to $\sim_{\Psi_\rho}$ in a state is denoted as $\lceil \sim_{\Psi_\rho} \rceil$. Let S_{Ψ_ρ} be a state which has the set $\lceil \sim_{\Psi_\rho} \rceil$ as its universe. For two equivalence classes $\lceil x \rceil_{\Psi_\rho}$ and $\lceil y \rceil_{\Psi_\rho}$ in $\lceil \sim_{\Psi_\rho} \rceil$, we define that

$$f^{S_{\Psi_\rho}}(\lceil x \rceil_{\Psi_\rho}) = \lceil y \rceil_{\Psi_\rho} \equiv \text{there exist } x \in \lceil x \rceil_{\Psi_\rho} \text{ and } y \in \lceil y \rceil_{\Psi_\rho} \text{ such that } (f(x) = y) \in \Psi_\rho.$$

The function f^{S_Ψ} is well-defined and total the same as discussed in [47]. That is, by the derivable property of functions (i.e., $(f(x_1) = y_1 \wedge f(x_2) = y_2 \wedge x_1 = x_2) \Rightarrow y_1 = y_2$) and Statement (2) of Lemma 5.5.1 that Ψ is deductively closed, we have,

$$(f^{S_{\Psi_\rho}}(\lceil x_1 \rceil_{\Psi_\rho}) = \lceil y_1 \rceil_{\Psi_\rho} \wedge f^{S_{\Psi_\rho}}(\lceil x_2 \rceil_{\Psi_\rho}) = \lceil y_2 \rceil_{\Psi_\rho} \wedge \lceil x_1 \rceil_{\Psi_\rho} = \lceil x_2 \rceil_{\Psi_\rho}) \Rightarrow \lceil y_1 \rceil_{\Psi_\rho} = \lceil y_2 \rceil_{\Psi_\rho}.$$

Furthermore, as Ψ_ρ contains witness, it means that, for each variable x, we have $f(x) = x_h$ in Ψ_ρ with a Henkin constant x_h. In this way, the function $f^{S_{\Psi_\rho}}$ can be treated as being total.

In order to distinguish from the previous case where only pure terms are considered, we use $\widetilde{\zeta}_\rho$ for a variable assignment which assigns an equivalence class $\lceil x \rceil_{\Psi_\rho}$ to a variable x, $[\![t]\!]_{\Psi_\rho,\widetilde{\zeta}_\rho}$ and $[\![\varphi]\!]_{\Psi_\rho,\widetilde{\zeta}_\rho}$ for the interpretations of a term t and a formula φ in state S_{Ψ_ρ} under $\widetilde{\zeta}_\rho$, respectively. For a ρ-term $t'(\overline{y}) \equiv \rho_{\overline{x}}(t|\varphi(\overline{x}, \overline{y}))$, the following property holds.

$$[\![t'(\overline{y})]\!]_{\Psi_\rho,\widetilde{\zeta}_\rho} = \lceil x \rceil_{\Psi_\rho} \equiv (t'(\overline{y}) = x) \in \Psi_\rho$$

Therefore, the set of equivalence classes of terms that may contain ρ-terms is also isomorphic to the set of equivalence classes of variables.

Same as the way discussed before, the interpretation of functions $\leq$ and $+$, and the interpretation of formulae $\Delta(f, \overline{t}, t_0)$, $\ddot{\Delta}(f, \overline{t}, t_0, t')$, $\in^3_\Delta (\Delta, f, \overline{t}, t_0)$ and $\in^4_{\ddot{\Delta}} (\ddot{\Delta}, f, \overline{t}, t_0, t')$ are fixed in the state S_{Ψ_ρ} under $\widetilde{\zeta}_\rho$.

Lemma 5.5.4. *Let φ be a formula. For any maximal consistent set Ψ_ρ of formulae containing witnesses, the following two statements are true:*

(a). *If $\varphi \in \Psi_\rho$, then $[\![\varphi]\!]_{\Psi_\rho,\tilde{\zeta}_\rho} = true$.*

(b). *If $\neg\varphi \in \Psi_\rho$, then $[\![\varphi]\!]_{\Psi_\rho,\tilde{\zeta}_\rho} = false$.*

Proof. The proof for this lemma is similar to the proof for Lemma 5.5.3. As the counterpart of Statement § in Lemma 5.5.3 and the cases of formulae $[\Delta]\varphi$, $\mathrm{upd}(r, \Delta)$ and $\mathrm{upm}(r, \ddot{\Delta})$ can be proven straightforwardly by following the ideas in Lemma 5.5.3, we skip the tedious details here.

□

Theorem 5.5.1. *Let M be a DB-ASM and Φ be a set of sentences. If $\Phi \models_M \varphi$, then $\Phi \vdash_M \varphi$.*

Proof. Assume that a formula φ is implied by Φ but is not derivable from Φ (i.e., $\Phi \models_M \varphi$ and $\Phi \nvdash_M \varphi$). By Lemma 5.5.1, there is a maximal consistent set Ψ of formulae containing witnesses such that $\Phi \cup \{\neg\varphi\} \subseteq \Psi$. According to Lemma 5.5.4, there is a state S_{Ψ_ρ} in which $[\![[\Delta]\varphi]\!]_{\Psi_\rho,\tilde{\zeta}_\rho} = false$. Therefore, φ is not implied by Φ (i.e., $\Phi \not\models_M \varphi$). This contradicts with the assumption that $\Phi \models_M \varphi$. Thus, the proof completes.

□

5.5.3 Definitional Extension of FO Logic

As DB-ASM is a variation of hierarchical ASM, which does not have recursive rule declarations, in this subsection, we adopt G.R.Renardel de Lavalette's approach to prove the completeness of $\mathbb{L}_M^{FO}$. This approach was also used by Stärk and Nanchen in [47] to prove the completeness of the logic for hierarchical ASMs.

Basically, the idea is to show that $\mathbb{L}_M^{FO}$ is a *definitional extension* of the first-order logic by translating formulae φ of $\mathbb{L}_M^{FO}$ into formulae φ' of the first-order logic, in which the following two properties hold:

1. $\varphi \Leftrightarrow \varphi'$ is derivable in $\mathbb{L}_M^{FO}$, and

2. if φ is derivable in $\mathbb{L}_M^{FO}$, then φ' is derivable in the first-order logic.

Therefore, we need to show that all the formulae in $\mathbb{L}_M^{FO}$ which are not first-order formulae can be translated into first-order formulae based on derivable equivalence in $\mathbb{L}_M^{FO}$.

First of all, the following principles in [134] for obtaining the general atomic formulae are applicable in our translation.

$$t_1 = t_2 \Leftrightarrow \exists x.(t_1 = x \wedge t_2 = x) \tag{5.1}$$

$$f(\bar{t}) = t_0 \Leftrightarrow \exists \bar{x}, y.(\bar{t} = \bar{x} \wedge t_0 = y \wedge f(\bar{x}) = y) \tag{5.2}$$

$$p(t_1, ..., t_n) \Leftrightarrow \exists x_1, ..., x_n.(t_1 = x_1 \wedge ... \wedge t_n = x_n \wedge p(x_1, ...x_n)) \tag{5.3}$$

$$\text{upd}(r, \Delta) \Leftrightarrow \exists X^3.(X^3 = \Delta \wedge \text{upd}(r, X^3)) \tag{5.4}$$

$$\text{upm}(r, \ddot{\Delta}) \Leftrightarrow \exists X^4.(X^4 = \ddot{\Delta} \wedge \text{upm}(r, X^4)) \tag{5.5}$$

$$\Delta(f, \bar{t}, t_0) \Leftrightarrow \exists \bar{x}, y.(\bar{x} = \bar{t} \wedge y = t_0 \wedge \Delta(f, \bar{x}, y)) \tag{5.6}$$

$$\ddot{\Delta}(f, \bar{t}, t_0, t') \Leftrightarrow \exists \bar{x}, y, z.(\bar{x} = \bar{t} \wedge y = t_0 \wedge z = t' \wedge \ddot{\Delta}(f, \bar{x}, y, z)) \tag{5.7}$$

It was shown in Subsection 5.5.1 that the second-order formulae used in $\mathbb{L}_M^{FO}$ can be translated into first-order formulae. Therefore, atomic formulae in the definitional extension of the first-order logic after the translation can be restricted to be the following:

- $x = y$
- $f(\bar{x}) = y$
- $P(x_1, ...x_n)$
- $\text{upd}(r, X^3_\Delta)$
- $\text{upm}(r, X^4_{\ddot{\Delta}})$
- $\in^3_\Delta (X^3, x_1, ..., x_3)$
- $\in^4_{\ddot{\Delta}} (X^4, x_1, ..., x_4)$

With respect to the extended formulae in $\mathbb{L}_M^{FO}$, we still need to eliminate the formulae def(r) and $[\Delta]\varphi$. As DB-ASM rules have no recursive definitions, the presence of def(r) is trivial. It means that def(r) can be replaced with the constant *true* as proposed in [134]. For the modal operator $[\Delta]$ in the formula $[\Delta]\varphi$, we use the following equivalences which are derivable from $\mathbb{L}_M^{FO}$.

$$[\Delta]x = y \Leftrightarrow (\text{con}(\Delta) \Rightarrow x = y) \tag{5.8}$$

$$[\Delta]f(\overline{x}) = y \Leftrightarrow (\text{con}(\Delta) \Rightarrow \Delta(f, \overline{x}, y)) \vee (\text{inv}(\Delta, f, \overline{x}) \wedge f(\overline{x}) = y) \tag{5.9}$$

$$[\Delta]p(x_1, ..., x_n) \Leftrightarrow (\text{con}(\Delta) \Rightarrow p(x_1, ..., x_n)) \tag{5.10}$$

$$[\Delta]\neg\varphi \Leftrightarrow (\text{con}(\Delta) \Rightarrow \neg[\Delta]\varphi) \tag{5.11}$$

$$[\Delta](\varphi \wedge \psi) \Leftrightarrow ([\Delta]\varphi \wedge [\Delta]\psi) \tag{5.12}$$

$$[\Delta](\varphi \vee \psi) \Leftrightarrow ([\Delta]\varphi \vee [\Delta]\psi) \tag{5.13}$$

$$[\Delta](\varphi \Rightarrow \psi) \Leftrightarrow ([\Delta]\varphi \Rightarrow [\Delta]\psi) \tag{5.14}$$

$$[\Delta]\forall x.\varphi \Leftrightarrow \forall x.[\Delta]\varphi \tag{5.15}$$

$$[\Delta]\exists x.\varphi \Leftrightarrow \exists x.[\Delta]\varphi \tag{5.16}$$

There is one thing worth to be mentioned. According to the semantics of formula $[\Delta]\varphi$ defined in Definition 5.2.2, $[\![[\Delta]\varphi]\!]_{S,\zeta} = true$ if $[\![\varphi]\!]_{S+\Delta,\zeta} = true$ for each state after applying a consistent Δ. Thus the following formalisation for the equivalent formula of $[\Delta]x = y$ would not be correct.

$$[\Delta]x = y \Leftrightarrow (\text{con}(\Delta) \wedge x = y).$$

When Δ is inconsistent, $[\Delta]x = y$ should be interpreted as $true$ because of no successor states. However, the above incorrect formalisation says that $[\Delta]x = y$ should be interpreted as $false$ when Δ is inconsistent. The same reason applies for the formalisation of the equivalent formulae of $[\Delta]f(\overline{x}) = y$, $[\Delta]p(x_1, ..., x_n)$ and $[\Delta]\neg\varphi$.

The following Theorems 5.5.2 and 5.5.3 about the completeness and compactness of $\mathbb{L}_M^{FO}$ are straightforward results following the properties of the first-order logic.

Theorem 5.5.2. *Let M be a DB-ASM and Φ be a set of sentences. If $\Phi \models_M \varphi$, then $\Phi \vdash_M \varphi$.*

Theorem 5.5.3. *If each finite subset of a set Φ of formulae is satisfiable, then Φ is satisfiable.*

Chapter 6

Partial Database Updates

In this chapter, we investigate the partial update problem in the context of complex-value databases. In database transformations, bounded parallelism is intrinsic and complex data structures form the core of each data model. Thus, the problem of partial updates arises naturally.

In Section 6.1, we first provide an example to illustrate the partial update problem encountered in a nested relational database which is built upon the NRDM discussed in Chapter 4. Then, we extend the discussion to the partial update problem in association with various type constructors used in a data model of complex-value databases.

The applicative algebra proposed by Gurevich and Tillmann [87, 88, 89] which addresses the partial update problem in a general setting is recalled in Section 6.2. However, using applicative algebras is not as smooth as its simple definition suggests. The root of the difficulties stems from the fact that it is common in data models to permit the arbitrary nesting of complex-value constructors. That is, an object in a complex-value database may be of a type built upon intricate nesting of a variety of type constructors over base types. Consequently, we need particles for each position in a complex value, and each nested structure requires its own parallel composition operation. Therefore, we have to deal with the theoretical possibility of infinitely many applicative algebras, which requires a mechanism for the construction of such algebras out of algebras for parts of the type of every object in a complex-value database. This leads to the question of how to efficiently check consistency for sets of partial updates.

In view of these problems, we propose an alternative solution to the problem of partial updates. The preliminaries such as the definition of partial locations, partial updates, and different kinds of dependencies among partial locations are handled in Section 6.3. In order to reflect a natural and flexible computing environment for database transformations, we relax the disjointness assumption on the notion of location. While, in principle, the locations bound to complex values are not independent from each other, we may consider each position within a complex value as a *sublocation*, which for simplicity of terminology we prefer to call also location. Then a partial update to a location is in fact a (partial) update to a sublocation.

In doing so, we can transform the problems of consistency checking and parallel composition into two stages: normalisation of shared updates and integration of total updates, which are discussed in Section 6.4 and Section 6.5, correspondingly. The first stage deals with compatibility of operators in shared updates and the second one deals with compatibility of clusters of exclusive updates.

We discuss the applications of partial updates in aggregate computing in Section 6.6.

6.1 The Problems

This section is to illustrate several typical partial update problems that may happen in complex-value databases. We begin with modifications on tuples in a relation since tuples represent a common view for locations in the relational model. As will be revealed in the following example, parallel manipulations on distinct attributes of a tuple are prohibited if only tuples are permissible locations in a state.

Example 6.1.1. Let S be a state containing a nested relation schema $R = \{A_1 : \{A_{11} : D_{11}, A_{12} : D_{22}\}, A_2 : D_2, A_3 : D_3\}$ and a nested relation $I(R)$ over R as shown in Figure 6.1. Then we have the locations $R(\{(a_{11}, a_{12}), (a'_{11}, a_{12})\}, b, c_1)$, $R(\{(a_{21}, a_{22}), (a'_{21}, a_{22})\}, b, c_2)$ and $R(\{(a_{31}, a_{32})\}, b_3, c_3)$, such that

- $val_S(R(\{(a_{11}, a_{12}), (a'_{11}, a_{12})\}, b, c_1)) = true$
- $val_S(R(\{(a_{21}, a_{22}), (a'_{21}, a_{22})\}, b, c_2)) = true$
- $val_S(R(\{(a_{31}, a_{32})\}, b_3, c_3)) = true$

Suppose that two update manipulations execute in parallel to modify values of attributes A_2 and A_3 in the same tuple. For example, the right rule in Figure 6.1 changes the attribute value b_3 in the third tuple to b. Meanwhile, the left rule in Figure 6.1 changes the attribute value c_3 in the same tuple to c_2. The left and right rules

$I(R)$

A_1		A_2	A_3
A_{11}	A_{12}		
$\{(a_{11},$	$a_{12}),$	b	c_1
$(a'_{11},$	$a_{12})\}$		
$\{(a_{21},$	$a_{22}),$	b	c_2
$(a'_{21},$	$a_{22})\}$		
$\{(a_{31},$	$a_{32})\}$	b_3	c_3

forall x, y, z **with** $R(x, y, z) \wedge y = b_3$
do
  par
    $R(x, y, z) := false$
    $R(x, y, c_2) := true$
  endpar
enddo

forall x, y, z **with** $R(x, y, z) \wedge y = b_3$
do
  par
    $R(x, y, z) := false$
    $R(x, b, z) := true$
  endpar
enddo

Figure 6.1: A problem of partial updates on attributes of a nested relation

yield pairs of updates $\{(R(\{[a_{31}, a_{32})\}, b_3, c_3)$, false), $(R(\{(a_{31}, a_{32})\}, b_3, c_2)$, true)$\}$ and $\{(R(\{(a_{31}, a_{32})\}, b, c_3)$, true), $(R(\{(a_{31}, a_{32})\}, b_3, c_3)$, false)$\}$, respectively. Since these two update manipulation are running in parallel, after the consistency checking and parallel composition, we get an update set $\Delta = \{(R(\{(a_{31}, a_{32})\}, b, c_3)$, true), $(R(\{(a_{31}, a_{32})\}, b_3, c_3)$, false), $(R(\{(a_{31}, a_{32})\}, b_3, c_2)$, true)$\}$.

However, applying the update set Δ in state S results in replacing the tuple $R(\{(a_{31}, a_{32})\}, b_3, c_3)$ by two tuples $R(\{(a_{31}, a_{32})\}, b, c_3)$, $R(\{(a_{31}, a_{32})\}, b_3, c_2)$ rather than a single tuple $R(\{(a_{31}, a_{32})\}, b, c_2)$ as expected in the resulting relation.

□

In order to solve this partial update problem on attributes, a straightforward solution is to add a finite number of attribute functions as locations for accessing attributes of tuples. Thus, in a state of relational database transformations, locations are extended to either an n-ary relational function symbol R with n arguments such as $R(a_1, ..., a_n)$, or a unary attribute function symbol with an argument in the form of $f_{R.A_1....A_k}(o)$ for a relation name

R, attributes $A_1, \ldots, A_k$ and an identifier o. Note that, the addition of attribute functions cannot replace the existence of relational functions. To delete a tuple from or add a tuple into a relation, we must still use relational functions. Attribute functions can only be used to modify the values of attributes, including NULL values. For the same reason, both relational functions and attribute functions are needed for a subrelation residing in a relation-valued attribute.

The following example illustrates how values of distinct attributes in the same tuple can be modified in parallel by using this approach.

Example 6.1.2. Let us consider again the nested relation $I(R)$ in Figure 6.2. Assume that o_i $(i = 1, 2, 3)$ are the identifiers of the tuples in the relation $I(R)$ and o_{ij} $(j = 1, 2)$ are the identifers of the nested tuples in the subrelations in the attribute A_1 of tuples o_i, respectively. Furthermore, we assume that there is a set of attribute functions with a one-to-one corresponding to the set $\mathcal{A}(R) = \{A_1, A_1.A_{11}, A_1.A_{12}, A_2, A_3\}$ of attributes contained in R, i.e., for each $A_k \in \mathcal{A}(R)$, $f_{R.A_k}(x) = y$ for a tuple identifier x in $I(R)$ and a value y in the domain of A_k. More concretely, the interpretation of attribute functions over $I(R)$ in state S is as follows.

- $val_S(f_{R.A_1}(o_1)) = \{(a_{11}, a_{12}), (a_{11}^{'}, a_{12})\}$
  $val_S(f_{R.A_1}(o_2)) = \{(a_{21}, a_{22}), (a_{21}^{'}, a_{22})\}$
  $val_S(f_{R.A_1}(o_3)) = \{(a_{31}, a_{32})\}$

- $val_S(f_{R.A_2}(o_1)) = b \quad val_S(f_{R.A_2}(o_2)) = b \quad val_S(f_{R.A_2}(o_3)) = b_3$

- $val_S(f_{R.A_3}(o_1)) = c_1 \quad val_S(f_{R.A_3}(o_2)) = c_2 \quad val_S(f_{R.A_3}(o_3)) = c_3$

- $val_S(f_{R.A_1.A_{11}}(o_{11})) = a_{11} \quad val_S(f_{R.A_1.A_{11}}(o_{12})) = a_{11}^{'}$
  $val_S(f_{R.A_1.A_{11}}(o_{21})) = a_{21} \quad val_S(f_{R.A_1.A_{11}}(o_{22})) = a_{21}^{'}$
  $val_S(f_{R.A_1.A_{11}}(o_{31})) = a_{31}$

- $val_S(f_{R.A_1.A_{12}}(o_{11})) = a_{12} \quad val_S(f_{R.A_1.A_{12}}(o_{12})) = a_{12}$
  $val_S(f_{R.A_1.A_{12}}(o_{21})) = a_{22} \quad val_S(f_{R.A_1.A_{12}}(o_{22})) = a_{22}$
  $val_S(f_{R.A_1.A_{12}}(o_{31})) = a_{32}$

Apart from the relational functions presented in Example 6.1.1, we also need the following relational functions referring to the sub-tuples in the relation $I(R)$.

$I(R)$

		A_1		A_2	A_3
		A_{11}	A_{12}		
o_1	o_{11}	$\{(a_{11},$	$a_{12}),$	b	c_1
	o_{12}	$(a'_{11},$	$a_{12})\}$		
o_2	o_{21}	$\{(a_{21},$	$a_{22}),$	b	c_2
	o_{22}	$(a'_{21},$	$a_{22})\}$		
o_3	o_{31}	$\{(a_{31},$	$a_{32})\}$	b_3	c_3

forall x **with** $R(x) \wedge f_{R.A_2}(x) = b_3$ **do**
  par
    $f_{R.A_2}(x) := b$
    $f_{R.A_3}(x) := c_2$
  endpar
enddo

Figure 6.2: A program updating two attributes in parallel

- $val_S(f_{R'.A_1}(o_1)(a_{11}, a_{12})) = true$
  $val_S(f_{R'.A_1}(o_1)(a'_{11}, a_{12})) = true$
  $val_S(f_{R'.A_1}(o_2)(a_{21}, a_{22})) = true$
  $val_S(f_{R'.A_1}(o_2)(a'_{21}, a_{22})) = true$
  $val_S(f_{R'.A_1}(o_3)(a_{31}, a_{32})) = true$

Using this approach, the rule in Figure 6.2 is able to modify values of attributes A_2 and A_3 of the same tuple in parallel.

□

The nested relation is just one example of complex-value databases. Other complex-value data models are possible by allowing arbitrary nesting of further type constructors such as list and multiset over base domains of a database transformation. The general idea is to similarly extend locations for each type constructor to be considered. Next we propose the locations necessary for two other common type constructors: list and multiset.

To change attribute values that are lists of elements, there are several approaches.

Following the terminology in [89], we call a *position* of a list as the number referring to an element of the list, and a *place* of a list to be before the first element, between two adjacent elements or after the last element of the list, which both start from zero and counts from left-to-right. Let us take the list $[b_{11}, b_{12}]$ as an example. There are three positions in $[b_{11}, b_{12}]$, where the element b_{11} is in position 0 and the element b_{12} is in position 1. Moreover, the list $[b_{11}, b_{12}]$ has three places, where place 0 is just before the element b_{11}, place 1 is between the elements b_{11} and b_{12} and place 2 is after the element b_{12}. In our discussion, we prefer to consider that, for a finite list s with length n, the locations of s are in the form of $f_s(k, k)$ for $k = 0, \ldots, n$ and $f_s(k, k+1)$ for $k = 0, \ldots, n-1$. That is, a location $f_s(k, k)$ indicates an insertion point of the list s, while a location $f_s(k, k+1)$ targets an element in the list to be modified or deleted. The symbol $\downarrow$ is used to indicate a deletion operation.

Similarly, there are also various ways to add locations for accessing attribute values which are multisets of elements. Here, we define multisets by associating a multiset with a pair (D, f), where D is a domain of elements and $f : D \to \mathbb{N}$ is a function from D to the set of natural numbers. Correspondingly, the locations referring to elements of a multiset $\mathcal{M}$ can be expressed as unary functions in the form of $f_{\mathcal{M}}(x)$ and an update $(f_{\mathcal{M}}(x), y)$ specifies that the element x has the number y of occurrence in the multiset $\mathcal{M}$. If y is zero, i.e., $(f_{\mathcal{M}}(x), 0)$, then we say that the element x does not exist in the multiset $\mathcal{M}$. In other words, the update says to remove x from the multiset $\mathcal{M}$. The following example illustrates how to specify updates to lists and multisets using such locations.

Example 6.1.3. Let $\mathcal{N}(D_2)$ be the set of all finite lists over the domain D_2 and, similarly, $\mathcal{M}(D_3)$ be the set of all finite multisets over the domain D_3. Suppose that we modify the relation $I(R)$ in Figure 6.2 to be a relation $I(R')$ in Figure 6.3 by associating the attribute A_2 with the domain $\mathcal{N}(D_2)$ and associating the attribute A_3 with the domain $\mathcal{M}(D_3)$. Then, the attribute functions for attributes A_2 and A_3 need to be changed correspondingly. That is,

- $val_S(f_{R'.A_2}(o_1)) = [b_{11}, b_{12}]$
- $val_S(f_{R'.A_2}(o_2)) = [b_{21}, b_{22}, b_{23}]$
- $val_S(f_{R'.A_2}(o_3)) = [b_3]$
- $val_S(f_{R'.A_3}(o_1)) = \{\!\{c_1, c_1\}\!\}$

$I(R')$

		A_1		A_2	A_3
		A_{11}	A_{12}		
o_1	o_{11}	$\{(a_{11},$	$a_{12}),$	$[b_{11}, b_{12}]$	$\{\!\{c_1, c_1\}\!\}$
	o_{12}	$(a'_{11},$	$a_{12})\}$		
o_2	o_{21}	$\{(a_{21},$	$a_{22}),$	$[b_{21}, b_{22}, b_{23}]$	$\{\!\{c_{21}, c_{22}, c_{22}\}\!\}$
	o_{22}	$(a'_{21},$	$a_{22})\}$		
o_3	o_{31}	$\{(a_{31},$	$a_{32})\}$	$[b_3]$	$\{\!\{c_{31}, c_{32}, c_{33}\}\!\}$

Figure 6.3: A relation in complex-value databases

- $vals_S(f_{R'.A_3}(o_2)) = \{\!\{c_{21}, c_{22}, c_{22}\}\!\}$
- $vals_S(f_{R'.A_3}(o_3)) = \{\!\{c_{31}, c_{32}, c_{33}\}\!\}$

With respect to the value $[b_{11}, b_{12}]$ of attribute A_2 in the tuple o_1, we may have a set of the locations $\{f_{R'.A_2}(o_1)(k,k) | k = 0, 1, 2\} \cup \{f_{R'.A_2}(o_1)(k, k+1) | k = 0, 1\}$.

- To insert an element b'_{11} just before b_{11}, the update would be $(f_{R'.A_2}(o_1)(0,0), b'_{11})$.
- To replace the element b_{11} with b'_{11}, the update would be $(f_{R'.A_2}(o_1)(0,1), b'_{11})$.
- To delete the element b_{11}, the update would be $(f_{R'.A_2}(o_1)(0,1), \downarrow)$.

As discussed before, for the value $\{\!\{c_{21}, c_{22}, c_{22}\}\!\}$ of attribute A_3 in the tuple o_2, we may have at least two locations $f_{R'.A_3}(o_2)(c_{21})$ and $f_{R'.A_3}(o_1)(c_{22})$.

- To add a new element c_{23} with the multiplicity 3, the update would be $(f_{R'.A_3}(o_2)(c_{23}), 3)$.
- To modify the multiplicity of the element c_{22} to be 1, the update would be $(f_{R'.A_3}(o_2)(c_{22}), 1)$.
- To delete the element c_{22} from the multiset, the update would be $(f_{R'.A_3}(o_2)(c_{22}), 0)$.

In addition, we may want to increase the multiplicity of an element in a multiset by a number k. In this case, we do not care about the original number of occurrence as long as

par
  forall x **with** $R(x) \wedge f_{R.A_2}(x) = b_3$ **do**
    $f_{R.A_2}(x) := b$
  enddo
  forall x, y, z **with** $R(x, y, z) \wedge y = b_3$ **do**
    $R(x, y, z) := false$
  enddo
endpar

Figure 6.4: A problem of partial updates after adding attribute functions

the multiplicity of the element has been increased by k. For this kind of modification, it would be natural to associate an additional operator with an update so as to describe how the multiplicity will be changed, for example, increase or decrease. □

The above approach of adding attribute functions works quite well in resolving the problem of partial updates on distinct attributes of a tuple or a subtuple. Nevertheless, the co-existence of locations $R(a_1, ..., a_n)$ for relational functions and $f_{R.A_1....A_k}(o)$ for attribute functions give rise to new problems as illustrated by the following example.

Example 6.1.4. In Figure 6.4 we have another rule executing over the nested relation $I(R)$ shown in Figure 6.2. Then the rule yields an update set containing two updates $(f_{R.A_2}(o_3), b)$ and $(R(\{(a_{31}, a_{32})\}, b_3, c_3)$, false). By using the standard definition for a consistent update set, we know that this update set is consistent. However, they are actually conflicting each other: The update $(f_{R.A_2}(o_3), b)$ intends to change the value of attribute A_2 of the tuple with identifier o_3 to b, while the update $(R(\{(a_{31}, a_{32})\}, b_3, c_3)$, false) intends to delete this tuple.

□

6.2 Applicative Algebra

In this section, we first recall the solution for the partial update problem proposed by Gurevich and Tillmann [87, 88, 89]. Then, we discuss the limitations of applying their solution in the context of complex-value databases.

6.2.1 General Framework

In [89], Gurevich and Tillmann have defined applicative algebra as a general algebraic framework for partial updates.

Definition 6.2.1. An *applicative algebra* $\mathfrak{A}$ consists of

- *elements*, which comprise a *trivial element* λ and a non-empty set of elements of a type τ,
- a monoid of total unary operations (called *particles*) over τ with functional composition and the identity operation id, including a *trivial particle* λ, and
  - $f(\lambda) = \lambda$ for every particle f, and
  - $\lambda(x) = \lambda$ for every element x.
- a *parallel composition* operation Ω, which assigns a particle $\Omega\mathcal{M}$ to each finite multiset $\mathcal{M}$ of particles, such that the following conditions are satisfied:
  - $\Omega\{\!\{f\}\!\} = f$,
  - $\Omega(\mathcal{M} \uplus \{\!\{id\}\!\}) = \Omega\mathcal{M}$, and
  - $\Omega(\mathcal{M} \uplus \{\!\{\lambda\}\!\}) = \lambda$.

As particles may come in families, a particle f of a family F is given by a tuple $(y_1, ..., y_n)$ of parameters which is called the *control* of f in F, and denoted by $f[y_1, ..., y_n](x)$.

Definition 6.2.2. A multiset $\mathcal{M}$ of particles is *consistent* if $\Omega\mathcal{M} \neq \lambda$.

Applicative algebra describes a collection of parallel modifications over elements of a type τ. In Gurevich's and Tillmann's work [87, 89], a partial update is a pair (ℓ, f) of a location ℓ of some type τ and a particle f over τ. For convenience, we say that such a partial update is of type τ. Therefore, an applicative algebra over a type τ becomes a general framework for handling all possible particles which may occur in partial updates of type τ. As this framework leaves the definition of $\Omega\mathcal{M} = \lambda$ open, it is empowered with full generality.

By characterising the algebraic properties of $\Omega\mathcal{M}$, different classes of applicative algebras can be identified. There is an old framework developed by Gurevich and Tillmann in [87], targeted for partial updates in the cases of types *counter*, *set* and *map*. As discussed in [88, 89], this old framework can be regarded as a special class of applicative algebras, called functional applicative algebra.

Definition 6.2.3. For any two particles f_1 and f_2 over the same type, f_1 and f_2 *commute* if $(f_1 \circ f_2)(x) = (f_2 \circ f_1)(x)$ for all x of that type, and f_1 and f_2 *malcommute* if $(f_1 \circ f_2)(x) \neq (f_2 \circ f_1)(x)$ for all x of that type.

Definition 6.2.4. A *functional applicative algebra* is an applicative algebra where

$$\Omega\{\!\{f_1, \dots, f_n\}\!\} = \begin{cases} f_1 \circ \cdots \circ f_n & \text{if all } f_i, f_j \text{ commute } (1 \leq i \neq j \leq n), \\ \lambda & \text{otherwise.} \end{cases}$$

A functional applicative algebra is called *apt* if every nontrivial particle maps nontrivial elements to nontrivial elements, and every nontrivial particles f_1, f_2 either commute or malcommute.

Lemma 6.2.1. *(Gurevich and Tillmann [87]). Let $\mathcal{M}$ be a multiset of particles of an apt functional applicative algebra. Then $\mathcal{M}$ is consistent iff every two members of $\mathcal{M}$ commute.*

Intuitively, a functional applicative algebra states that a multiset of particles of a type τ is consistent if the functional composition of them is order-independent. Although this approach was successfully applied on types *counter*, *set* and *map*, it has been pointed out in [88, 89] that a functional applicative algebra is too restrictive for particles of type *sequence*. More specifically, for two compatible modifications f_1, f_2 over type *sequence* such that $\Omega\{\!\{f_1, f_2\}\!\} \neq \lambda$, it is still possible that $f_1 \circ f_2 \neq f_2 \circ f_1$ holds, i.e., different orders of executing sequential compositions of particles leads to different results.

A solution for the partial update problem over type *sequence* in the setting of applicative algebra was proposed in [89]. Let $\mathfrak{A}$ be an applicative algebra over type τ, which contains at least two nontrivial elements. Then an applicative algebra $seq(\mathfrak{A})$ over type *sequence* is defined by:

- $seq(\mathfrak{A})$ has infinitely many nontrivial elements that are finite sequences over τ, together with the trivial sequence λ.
- $seq(\mathfrak{A})$ has three families of particles: *substitution*, *alteration* and *position* particles.
  - substitution particle $sub[a, s_1, s_2]$ for $a \in \mathbb{N}$, a sequence s_1 of unary predicates over $\mathfrak{A}$ and a nontrivial replacement sequence s_2;
  - alteration particle $alter[a, f]$ for $a \in \mathbb{N}$ and a particle f of $\mathfrak{A}$;

  - position particles $pos[a]$ for $a \in \mathbb{N}$.

- for a multiset $\mathcal{M}$ of particles, $\Omega\mathcal{M} = \lambda$ if either of the following conditions holds.

  - $\mathcal{M}$ contains particles f_1, f_2 such that neither $f_1 < f_2$ nor $f_1 > f_2$, and f_1 is a substitution and f_2 is either a substitution or an alteration particle.
  - $\mathcal{M}$ contains particles $alter[a, f_1'], \ldots, alter[a, f_n']$ with the same a such that $\Omega\{\!\{f_1', \ldots, f_n'\}\!\} = \lambda$ in $\mathfrak{A}$.

Let $\sharp s$ denote the length of a sequence s. Then a substitution particle $sub[a, s_1, s_2]$ is

- an insertion at a with s_2 if s_1 is empty,
- a deletion from a to $a + \sharp s_1 - 1$ if s_1 is not empty but s_2 is empty, or
- a replacement from a to $a + \sharp s_1 - 1$ with s_2 if both s_1 and s_2 are nonempty.

An alteration particle $alter[a, f]$ alters the element of $\mathfrak{A}$ at a with the particle f of $\mathfrak{A}$. A position particle $pos[a]$ checks whether the position a is present in a file. A particle f_1 is *to the left* of f_2, denoted as $f_1 < f_2$, if the part of a file that f_1 involves is to the left of the part of a file that f_2 involves.

6.2.2 Limitations

In the context of establishing applicative algebra, there is an important assumption that locations of partial updates are disjoint. Let $\mathcal{M}$ be a multiset of particles, $Loc(\mathcal{M})$ be the set of locations occurring in $\mathcal{M}$ and $\mathcal{M}_\ell$ be a submultiset of $\mathcal{M}$ such that $\mathcal{M}_\ell = \{\!\{(f, \ell) | (f, \ell) \in \mathcal{M}\}\!\}$. Then, by the disjointness condition of locations, we have,

$$\Omega\mathcal{M} \neq \lambda \text{ iff } \Omega\mathcal{M}_\ell \neq \lambda \text{ for each } \ell \in Loc(\mathcal{M}).$$

However, if we remove the assumption that locations in a state are disjoint, then $\Omega\mathcal{M}$ may be λ even when $\Omega\mathcal{M}_\ell \neq \lambda$ for each $\ell \in Loc(\mathcal{M})$. This has been exemplified in Example 6.1.4, where the location $R(\{(a_{31}, a_{32})\}, b_3, c_3)$ subsume the location $f_{R.A_2}(o_3)$ and the two updates $(f_{R.A_2}(o_3), b)$ and $(R(\{(a_{31}, a_{32})\}, b_3, c_3), \text{false})$ are not consistent. That is, due to the arbitrary nesting of complex-value constructors, locations of complex-value database are often defined at multiple abstraction levels and thereby non-disjoint. In fact, allowing locations of different abstraction levels plays a vital role in supporting the requests of updating complex-values of a database at different granularity.

This brings us to the question of how to utilise the approach of applicative algebra to solve the partial update problem in the setting of non-disjoint locations. A cursory investigation reveals two possible approaches:

- Straightforwardly, we can convert a multiset of partial updates that have non-disjoint locations into a multiset of partial updates that have disjoint locations. Then applicative algebras can be applied to different types in a similar way as suggested in [89].

- The second one is to establish a mechanism for the nested construction of applicative algebras out of applicative algebras in accordance with complex data structures used in the data models of complex-value databases.

In other words, the first approach suggests to transform updates with nested locations into updates with nested modifications but disjoint locations. Because all sorts of particles used to modify the nested internal structure of an element have to be defined on the outermost type of the element, this immediately leads to particles with complicated controls which are necessary to encode the nested modifications. Generally, the more complicated a type, the more complex the encoding of controls for particles. As a consequence of this, it is imperative to define a notion of the parallel composition operation for partial updates with the nested structures in their controls. Pragmatically, this turns out to be hard to implement in a complex-value database environment. The following example illustrates a possible way of encoding the nested modifications into the controls of particles.

Example 6.2.1. Recall the relation $I(R^{'})$ of the complex-value database provided in Figure 6.3. Assume that the set of disjoint locations chosen in this example is in the form of the relational function symbol $R^{'}$ with three arguments, and there are four updates $(f_{R^{'}.A_2}(o_3)(0,0), b_3^{'})$, $(f_{R^{'}.A_2}(o_3)(1,1), b_3^{''})$, $(f_{R^{'}.A_1.A_{11}}(o_{21}), a_{21}^{'})$ and $(f_{R^{'}.A_3}(o_3)(c_{31}), 2)$.

Now we need to convert these four updates into updates with the location $R^{'}(\{(a_{31}, a_{32})\}, [b_3], \{\!\{c_{31}, c_{32}, c_{33}\}\!\})$. To achieve this, we have to encode the parameters of locations into the controls of particles, such as,

- $f[R^{'}.A_2, (list\_insert(0, 0, b_3^{'}))](x)$
- $f[R^{'}.A_2, (list\_insert(1, 1, b_3^{''}))](x)$

- $f[R'.A_1.A_{11},(replace\_value(o_{21},a'_{21}))](x)$
- $f[R'.A_3,(modify\_multiset(c_{31},2))](x)$

where $list\_insert(x_1,x_2,y)$ inserts y into the place between x_1 and x_2 of a list, $replace\_value(x,y)$ replaces the specified attribute value in a tuple with identifier x with y, and $modify\_multiset(x,y)$ modifies the multiplicity of element x in a multiset to be y. □

Therefore, from an algebraic point of view, we still need to establish algebras out of algebras for handling the nested controls of particles in the first approach. This would be similar to but worse than directly establishing algebras out of algebras for locations because the notion of control of a particle is very confusing. How should we formalise particles with nested controls, in which elements from the base set of a database transformation might appear, without violating the abstract state and genericity postulates requested by a database transformation?

Now we look into the second approach. To help the illustration, we will still base the discussion on the relation $I(R')$ in Figure 6.3.

Example 6.2.2. We know that the relation $I(R')$ in Figure 6.3 has the following schema:

$$R' = \{A_1 : \{A_{11} : D_{11}, A_{12} : D_{22}\}, A_2 : \mathcal{N}(D_2), A_3 : \mathcal{M}(D_3)\},$$

and the associated type:

$$typ(R') = \{(A_1 : \{(A_{11} : D_{11}, A_{12} : D_{22})\}, A_2 : \mathcal{N}(D_2), A_3 : \mathcal{M}(D_3))\}.$$

Assume that $\mathfrak{A}_i$ $(i = 11, 12, 2, 3)$ are applicative algebras built upon the domains D_i, and we use the notations $set(\mathfrak{A})$, $tup(\mathfrak{A})$, $seq(\mathfrak{A})$ and $mul(\mathfrak{A})$ to denote applicative algebras built upon an applicative algebras $\mathfrak{A}$ for types *set*, *tuple*, *sequence* and *multiset*, respectively. Then by the type $typ(R')$, the construction for the following nested applicative algebra needs to be established.

$$set(tup(set(tup(\mathfrak{A}_{11},\mathfrak{A}_{12})),seq(\mathfrak{A}_2),mul(\mathfrak{A}_3)))$$

□

This kind of construction is quite complicated. Furthermore, there are several issues which need to be further explored.

- How to properly reflect the consistency and integration of partial updates at a particular level in the consistency and integration of partial updates at higher levels?

- Is there an efficient algorithm that can handle the consistency and integration of a multiset of partial updates at different abstraction levels?

In the rest of this chapter, we will develop a customised and efficient mechanism to handle the partial update problem in the context of complex-value databases.

6.3 Preliminaries

In order to deal with the problems discussed in the previous section, we first have to relax the standard notion of location, and then formally define partial updates.

6.3.1 Partial Locations

To avoid the confusion with the standard notion of location used in ASMs (see Definition 3.1.8), here we call such locations *prime locations* of a database transformation. While, in principle, prime locations bound to complex values are independent from each other, we may consider each position within a complex value as a non-prime location. For this, we need to add some auxiliary function symbols into the signature of a state.

Definition 6.3.1. Let S be a state of a database transformation, f be an auxiliary dynamic function symbol of arity n in the state signature and $a_1, ..., a_n$ be elements in the base set of S. Then $f(a_1, ..., a_n)$ is called a *non-prime location* of S.

Definition 6.3.2. A location ℓ_1 *subsumes* a location ℓ_2 (notation: $\ell_2 \sqsubseteq \ell_1$) if

- for all states S of a database transformation, $val_S(\ell_1)$ uniquely determines $val_S(\ell_2)$.

We call a location ℓ_2 the *sublocation* of a location ℓ_1 if $\ell_2 \sqsubseteq \ell_1$ holds. A location is the sublocation of itself. Furthermore, there is a *trivial sublocation* $\ell_\perp$ for every location.

Example 6.3.1. $f_{R'.A_2}(o_3)(0,0)$, $f_{R'.A_2}(o_3)(1,1)$ and $f_{R'.A_3}(o_3)(c_{31})$ discussed in Example 6.2.1 are the non-prime locations. Moreover, they are the sublocations of $R'(\{(a_{31}, a_{32})\}, [b_3], \{\!\{c_{31}, c_{32}, c_{33}\}\!\})$. □

From a constructive point of view, a prime location may be considered as an algebraic structure in which its sublocations refer to parts of the structure. Since such a structure is always constructed by using type constructors like set, tuple, list, multiset, etc. from a specific data model, we only allow sublocations of a prime location, which either subsume or disjoint one another, to be partial locations by the following definition. This restriction is more a technicality so that we can focus on discussing the integration and consistency checking of partial updates. Extending to the general case would be straightforward after adding a decomposition procedure to eliminate sublocations that are overlapping but do not subsume one another.

Example 6.3.2. For the prime location $\ell = R'(\{(a_{31}, a_{32})\}, [b_3], \{\!\{c_{31}, c_{32}, c_{33}\}\!\})$ in the relation $I(R')$ of Figure 6.3, it would be possible to consider $f_{R'.A_2}(o_3)(0, 2)$ as a sublocation of ℓ. However, we will treat it as a shortcut for a collection of partial locations, i.e., $\{f_{R'.A_2}(o_3)(0, 1), f_{R'.A_2}(o_3)(1, 1), f_{R'.A_2}(o_3)(1, 2)\}$, rather than a partial location itself. □

Before presenting the formal definition of partial locations, we need to define two operations $\sqcup_\ell$ and $\sqcap_\ell$. Let ℓ_1, ℓ_2, ℓ_3 be three (prime or non-prime) locations. Then $\ell_1 \sqcup_\ell \ell_2 = \ell_3$ if $\ell_1 \sqsubseteq \ell_3$, $\ell_2 \sqsubseteq \ell_3$ and there is no other $\ell \in \mathcal{L}_\ell$ such that $\ell \neq \ell_3$, $\ell_1 \sqsubseteq \ell$, $\ell_2 \sqsubseteq \ell$ and $\ell \sqsubseteq \ell_3$. The operation $\sqcap_\ell$ can be similarly defined, that is, $\ell_1 \sqcap_\ell \ell_2 = \ell_3$ if $\ell_3 \sqsubseteq \ell_1$, $\ell_3 \sqsubseteq \ell_2$ and there is no other $\ell \in \mathcal{L}_\ell$ such that $\ell \neq \ell_3$, $\ell \sqsubseteq \ell_1$, $\ell \sqsubseteq \ell_2$ and $\ell_3 \sqsubseteq \ell$. We say $\ell_1 \sqcap_\ell \ell_2 = \ell_\perp$ if ℓ_1 and ℓ_2 are disjoint, i.e., neither $\ell_1 \sqsubseteq \ell_2$ nor $\ell_2 \sqsubseteq \ell_1$ hold.

Definition 6.3.3. Let S be a state of a database transformation. Then the set of *partial locations* of S is the smallest set such that

- each prime location is a partial location, and
- if each prime location ℓ is an algebraic structure $(\mathcal{L}_\ell, \sqcup, \sqcap, \ell, \ell_\perp)$ satisfying the following conditions, then each sublocation of ℓ is a partial location.
  - $(\mathcal{L}_\ell, \sqcup_\ell, \sqcap_\ell)$ is a lattice, consisting of a set $\mathcal{L}_\ell$ of all sublocations of ℓ, and two binary operations $\sqcup_\ell$ (i.e., join) and $\sqcap_\ell$ (i.e., meet) on $\mathcal{L}_\ell$,
  - ℓ is the identity element for the join operation $\sqcup_\ell$,
  - $\ell_\perp$ is the identity element for the meet operation $\sqcap_\ell$, and
  - for any ℓ_1 and ℓ_2 in $\mathcal{L}_\ell$, one of the following conditions must be satisfied:
    * $\ell_1 \sqcup_\ell \ell_2 = \ell_1$,

* $\ell_1 \sqcup_\ell \ell_2 = \ell_2$,
* $\ell_1 \sqcap_\ell \ell_2 = \ell_\bot$.

Example 6.3.3. Let us consider the prime location $R'(\{(a_{31}, a_{32})\}, [b_3], \{\!\{c_{31}, c_{32}, c_{33}\}\!\})$ in the relation $I(R')$ of Figure 6.3. This prime location can be regarded as an algebraic structure illustrated in Figure 6.5, where the label i of a node in the picture at the top corresponds to the index i of the sublocation ℓ_i in the table at the bottom. All conditions required by Definition 6.3.3 are satisfied, therefore, these sublocations are partial locations.

□

In addition to the subsumption relation, one partial location may be dependent on another partial location, i.e., there exists the dependence relation over partial locations of a state.

Definition 6.3.4. A location ℓ_1 *depends* on a location ℓ_2 (notation: $\ell_2 \trianglelefteq \ell_1$) if $val_S(\ell_2) = \bot$ implies $val_S(\ell_1) = \bot$ for all states S.

The dependency relation $\trianglelefteq$ is said to be *strict* on the location ℓ, if, for all $\ell_1, \ell_2, \ell_3 \in \mathcal{L}_\ell = \{\ell' \mid \ell' \sqsubseteq \ell\}$, we have that whenever $\ell_1 \trianglelefteq \ell_2$ and $\ell_1 \trianglelefteq \ell_3$ hold, then either $\ell_2 \trianglelefteq \ell_3$ or $\ell_3 \trianglelefteq \ell_2$ holds as well.

However, such a dependency may also occur without nesting, the prominent examples being sequences and trees.

Example 6.3.4. Consider the partial locations $f_{R'.A_2}(o_3)(0,0)$, $f_{R'.A_2}(o_3)(0,1)$ and $f_{R'.A_2}(o_3)(1,1)$ in the relation $I(R')$ of Figure 6.3. As $f_{R'.A_2}(o_3)(k_1', k_2') \trianglelefteq f_{R'.A_2}(o_3)(k_1, k_2)$ holds for $k_1' < k_2$, the dependency relation $\trianglelefteq$ is strict on $f_{R'.A_2}(o_3)$.

□

B+-trees provide examples for non-strict dependency relations that are at the same time not induced by subsumption.

A partial location ℓ_2 that is subsumed by a partial location ℓ_1 certainly depends on it in the sense that if it is bound to a value other than $\bot$ (representing undefinedness), then also ℓ_1 cannot be bound to $\bot$. So the following lemma is straightforward.

Lemma 6.3.1. *For two partial locations ℓ_1, ℓ_2 with $\ell_2 \sqsubseteq \ell_1$ we also have $\ell_1 \trianglelefteq \ell_2$.*

Proof. Let S be a state. As $val_S(\ell_1)$ uniquely determines $val_S(\ell_2)$, clearly, $val_S(l_1) = \bot$ implies $val_S(l_2) = \bot$. That is, ℓ_2 *depends* on ℓ_1, i.e. $\ell_1 \trianglelefteq \ell_2$.

□

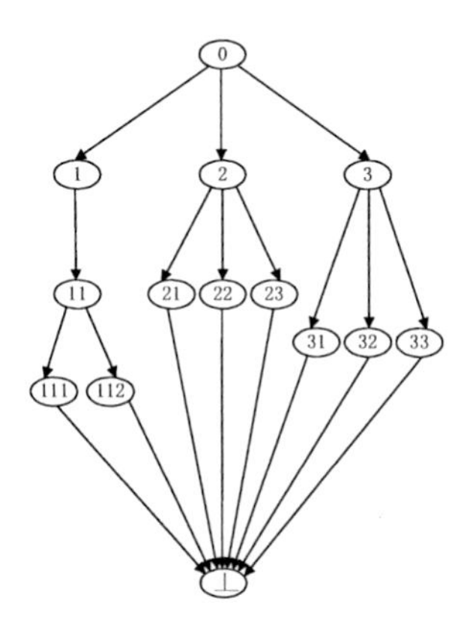

ℓ_0	$R'(\{(a_{31}, a_{32})\}, [b_3], \{\!\{c_{31}, c_{32}, c_{33}\}\!\})$
ℓ_1	$f_{R'.A_1}(o_3)$
ℓ_2	$f_{R'.A_2}(o_3)$
ℓ_3	$f_{R'.A_3}(o_3)$
ℓ_{11}	$f_{R'.A_1}(o_3)(a_{31}, a_{32})$
ℓ_{111}	$f_{R'.A_1.A_{11}}(o_{31})$
ℓ_{112}	$f_{R'.A_1.A_{12}}(o_{31})$
ℓ_{21}	$f_{R'.A_2}(o_3)(0,0)$
ℓ_{22}	$f_{R'.A_2}(o_3)(0,1)$
ℓ_{23}	$f_{R'.A_2}(o_3)(1,1)$
ℓ_{31}	$f_{R'.A_3}(o_3)(c_{31})$
ℓ_{32}	$f_{R'.A_3}(o_3)(c_{32})$
ℓ_{33}	$f_{R'.A_3}(o_3)(c_{33})$
$\ell_{\perp}$	

Figure 6.5: An algebraic structure of a prime location

6.3.2 Partial Updates

To formalise the definition of partial updates, we associate a type with each partial location $\ell = f(a_1, ..., a_n)$, such that the type $\tau(\ell)$ of $f(a_1, ..., a_n)$ is the codomain of the function $f : D_1 \times ... \times D_n \rightarrow D$, i.e., $\tau(\ell) = D$. Therefore, a type of partial locations can be a built-in type provided by database systems, such as *String*, *Int*, *Date*, etc., a complex-value type constructed by using type constructors in a data model, such as set, tuple, list and multiset constructors, or a customised type defined by users, i.e., user-defined types (UDTs) used in database applications.

Example 6.3.5. Reconsider the partial locations $\ell_1 = f_{R'.A_1}(o_3)$, $\ell_2 = f_{R'.A_2}(o_3)$ and $\ell_3 = f_{R'.A_3}(o_3)$ in Example 6.3.3 and the relation $I(R')$ in Figure 6.3. They have the following types:

- $\tau(\ell_1) = \mathcal{P}(\mathcal{NT}^2(D_{11}, D_{12}))$,
- $\tau(\ell_2) = \mathcal{N}(D_2)$, and
- $\tau(\ell_3) = \mathcal{M}(D_3)$,

where $\mathcal{P}(D)$, $\mathcal{N}(D)$ and $\mathcal{M}(D)$ denote the set of all subsets, lists and multisets over the domain D, and $\mathcal{NT}^2(D_1, D_2)$ denotes the set of all 2-ary tuples over the domains D_1 and D_2.

Instead of particles, we will formalise partial updates of a database transformation in terms of two types: exclusive and shared updates.

Definition 6.3.5. An *exclusive update* is a pair (ℓ, b) consisting of a location ℓ and a value b of the same type τ as ℓ. A *shared update* is a triple (ℓ, b, μ) consisting of

- a location ℓ of type τ,
- a value b of type τ, and
- a binary operator $\mu : \tau \times \tau \rightarrow \tau$.

For a state S and an update set Δ containing a single (exclusive or shared) update, we have

$$val_{S+\Delta}(\ell) = \begin{cases} b & \text{if } \Delta = \{(\ell, b)\} \\ \mu(val_S(\ell), b) & \text{if } \Delta = \{(\ell, b, \mu)\} \end{cases}$$

Although exclusive updates have the similar form to updates of ASMs defined in a standard way, exclusive updates are allowed to have partial locations. It means that the locations of two exclusive updates may have a dependency relationship, whereas the locations of two standard updates of ASMs are assumed to be independent. Therefore, the notion of exclusive update generalises the notion of update in ASMs. Updates defined in ASMs become exclusive updates to prime locations in our definition.

In a shared update (ℓ, b, μ), the binary operator μ is used to specify how the value b partially affects the content of ℓ in a state. When multiple partial updates are generated to the same location simultaneously, a multiset of partial updates is obtained. For example, a location ℓ of type $\mathbb{N}$ may associate with a multiset of shared updates $\{\!\{(\ell, 10, +), (\ell, 10, +), (\ell, 5, -)\}\!\}$ (i.e., increase the content of ℓ by 10 twice and decrease the content of ℓ by 5 once). Another reason for the introduction of a binary operator μ in a shared update is more specific to the requirements of database transformations. Let us elaborate this point by the following example.

Example 6.3.6. Suppose that we have two partial updates $(\ell, 5, +)$ and $(\ell, 1, \times)$, which may also be expressed by particles: $incr[5](x)$ and $times[1](x)$. Assume that we define $\Omega\mathcal{M}_\ell = \lambda$ iff the functional composition of particles in $\mathcal{M}$ is order-independent. In this case, as ℓ is a type of $\mathbb{N}$ (become a type *counter* if using the terminology of [89]), this definition of the parallel composition operation for particles associated with ℓ is quite reasonable.

According to the above definition, we may conclude that these two particles are consistent, and therefore two partial updates are consistent. However, an important principle of database transformations is the genericity principle which are stipulated by the abstract state and genericity postulates. Respecting these principles mean that, if two partial updates $(\ell, 5, +)$ and $(\ell, 1, \times)$ are consistent, then any two partial updates $(\ell, \varsigma(5), +)$ and $(\ell, \varsigma(1), \times)$ should also be consistent, where ς is a permutation over elements in base domains of a database transformation. Obviously, this is not correct. If we swap the elements 1 and 5, then $(\ell, 1, +)$ and $(\ell, 5, \times)$ are not consistent. □

The use of a binary operator μ in shared updates helps us to separate the concerns relating to database instance and database schema. By this separation, the consistency checking of incompatible operators can be conducted at a database schema level, which will be further discussed in the next section. So this viewpoint is efficient in practice, particularly for those database applications with large data-sets.

The border between shared and exclusive updates lies in whether the modifications are intended to share with each other. An exclusive update (ℓ, b) may be regarded as a special kind of shared update (ℓ, b, μ), where μ is a projection function on pairs of values of type τ. Nonetheless, we prefer to separate the treatment of exclusive and shared updates, as only the latter ones require considerable care for consistency checking of operators. Because of the separation, we provide different update rules to generate exclusive and shared updates.

Definition 6.3.6. Let t_1 and t_2 be terms of type τ, and μ be a binary operator over type τ. Then the *partial update rules* take one of the following two forms

- the rule for exclusive updates:

$$t_1 \leftleftarrows t_2;$$

- the rule for shared updates:

$$t_1 \leftleftarrows^{\mu} t_2.$$

Semantically, the partial update rules generate updates in a multiset. Let S be a state and ζ be a variable assignment. Then

- $\ddot{\Delta}(t_1 \leftleftarrows t_2, S, \zeta) = \{\!\{(\ell, b)\}\!\}$ and
- $\ddot{\Delta}(t_1 \leftleftarrows^{\mu} t_2, S, \zeta) = \{\!\{(\ell, b, \mu)\}\!\}$,

where $\ell = t_1[a_1/x_1, \ldots, a_n/x_n]$ for $var(t_1) = \{x_1, \ldots, x_n\}$ and $\zeta(x_i) = a_i$ $(i = 0, \ldots, n)$, and $val_{S,\zeta}(t_2) = b$.

Remark 6.3.1. The addition of auxiliary functions as locations of a state requires a shifted view for partial updates in our definition. In contrast to an update (ℓ, b) defined in standard ASMs, in which $val_{S+\{(\ell,b)\}}(\ell) = b$ holds for every state S, the partial updates considered here do not satisfy such a condition.

This can be illustrated by the following example.

Example 6.3.7. Consider a state S that contains the relation $I(R')$ in Figure 6.3 and the partial updates $(f_{R'.A_2}(o_3)(0,0), d_{31})$ and $(f_{R'.A_2}(o_3)(0,1), d_{32})$. Applying these partial updates will change the value of attribute A_2 at the tuple with identifier o_3 from $[d_3]$ in the state S to $[d_{31}, d_{32}]$ in the successor state $S' = S + \{(f_{R'.A_2}(o_3)(0,0), d_{31}), (f_{R'.A_2}(o_3)(0,1), d_{32})\}$. However, $val_{S'}(f_{R'.A_2}(o_3)(0,0)) \neq d_{31}$ and $val_{S'}(f_{R'.A_2}(o_3)(0,1)) \neq d_{32}$. Instead, we may say, $val_{S'}(f_{R'.A_2}(o_3)(0,0)) = null$ and $val_{S'}(f_{R'.A_2}(o_3)(0,1)) = d_{31}$.

For convenience, we will call partial location as location in the rest of this chapter.

6.4 Normalisation of Shared Updates

Normalisation of a multiset $\ddot{\Delta}$ of partial updates is the process of merging all shared updates to the same location into a single exclusive update. Thus, $\ddot{\Delta}$ is transformed into an update set Δ containing only exclusive updates.

Definition 6.4.1. An update multiset $\ddot{\Delta}$ is in the *normal form* if each update in it is an exclusive update with multiplicity 1.

As a convention, let $Loc(\ddot{\Delta})$ and $Opt(\ddot{\Delta})$ denote the set of locations and the set of operators occurring in an update multiset $\ddot{\Delta}$, respectively, i.e., $Loc(\ddot{\Delta}) = \{\ell | (\ell, b, \mu) \in \ddot{\Delta}\}$ and $Opt(\ddot{\Delta}) = \{\mu | (\ell, b, \mu) \in \ddot{\Delta}\}$, and $\ddot{\Delta}_\ell$ denotes the submultiset of an update multiset $\ddot{\Delta}$ containing all shared updates that have the location ℓ, i.e., $\ddot{\Delta}_\ell = \{\!\{u | u \in \ddot{\Delta} \text{ and } u = (\ell, b, \mu)\}\!\}$.

6.4.1 Operator-Compatibility

The notion of operator-compatible addresses the inconsistencies arising from shared updates to the same location in an update multiset, no matter which abstraction level their locations reside at and whether they are dependent on other locations in the same update multiset.

We begin with some intuitive operators coming from elementary arithmetic.

Example 6.4.1. Let $\mathbb{Q}^*$ be the set of rational numbers excluding zero and $\mathbb{R}$ be the set of real numbers. Then addition $+$ and substraction $-$ are operators over $\mathbb{R}$, and multiplication $\times$ and division $\div$ are operators over $\mathbb{Q}^*$, respectively. Suppose that ℓ is a location of type $\mathbb{Q}^*$, then the following modifications can be executed in parallel.

par
$\quad \ell \leftrightarrows^{+} b_1$
$\quad \ell \leftrightarrows^{-} b_2$
$\quad \ell \leftrightarrows^{\times} b_3$
$\quad \ell \leftrightarrows^{\div} b_4$
endpar

For this rule, the update multiset $\ddot{\Delta}_\ell = \{\!\{(\ell, b_1, +), (\ell, b_2, -), (\ell, b_3, \times), (\ell, b_4, \div)\}\!\}$ is obtained. The operators in the submultisets $\{\!\{(\ell, b_1, +), (\ell, b_2, -)\}\!\}$ and $\{\!\{(\ell, b_3, \times), (\ell, b_4, \div)\}\!\}$

are compatible. Nevertheless, the operators in $\ddot{\Delta}_\ell$ is not compatible, because applying updates in $\ddot{\Delta}_\ell$ in different orders yields different results. □

Many languages developed for database manipulations have set-theoretic operations, such as Structured Query Language (SQL), Relational Algebra (RA), etc. The partial-update problem relating to set-theoretic operations is about the parallel manipulations on sets via various set-based operations. The following example provides a simple scenario in which, after a main computation initialises a set of subcomputations, each of subcomputations may yield a set of values which is then put into the final result by using union operations in parallel.

Example 6.4.2. Let $\mathcal{P}(D)$ be the powerset of the domain D. Then the set-based operations: union $\cup$, intersection $\cap$, difference $-$, symmetric difference $\diamond$, etc. can be regarded as common operators over domain $\mathcal{P}(D)$. The following rule produces an operator-compatible update multiset $\{\!\{(\ell, \{b_1, b_2\}, \cup), (\ell, \{b_2, b_3, b_4\}, \cup)\}\!\}$.

par
$\ell \leftrightharpoons^{\cup} \{b_1, b_2\}$
$\ell \leftrightharpoons^{\cup} \{b_2, b_3, b_4\}$
endpar □

These examples motivate a straightforward definition of operator-compatibility in terms of order-independent application of shared updates to the same location, that is, we expect the same result regardless of the order in which shared updates in a given update multiset is applied.

Definition 6.4.2. Let $\ddot{\Delta}_\ell = \{\!\{(\ell, a_i, \mu_i) \mid i = 1, ..., k\}\!\}$ be a multiset of shared updates to the same location ℓ. Then $\ddot{\Delta}_\ell$ is *operator-compatible* if for any two permutations $(p_1, ..., p_k)$ and $(q_1, ..., q_k)$, we have for all x

$$\mu_{p_k}(...\mu_{p_1}(x, a_{p_1})..., a_{p_k}) = \mu_{q_k}(...\mu_{q_1}(x, a_{q_1})..., a_{q_k}).$$

An update multiset $\ddot{\Delta}$ is *operator-compatible* if $\ddot{\Delta}_\ell$ is operator-compatible for each $\ell \in Loc(\ddot{\Delta})$.

The notion of being order-independent in Definition 6.4.2 does not apply to the whole update multiset $\ddot{\Delta}$, as locations may depend on each other. Two shared updates to distinct but dependent locations may generate different results when applied in different orders. Furthermore, as illustrated in Example 6.4.1, the order-independence of operators is easy to check when the number of shared updates is small. However, in case of a large number of shared updates, compatibility checking by means of exploring all possible orderings is far too time-consuming. Therefore, we introduce an algebraic approach to characterise the operator-compatibility of shared updates to the same location.

Corollary 6.4.1. *An update multiset $\ddot{\Delta}_\ell$ is operator-compatible iff the order to execute any two updates in $\ddot{\Delta}_\ell$ is insignificant.*

The above corollary is a straightforward result from Definition 6.4.2. By this corollary, we will start with examining the operator-compatibility of two shared updates to the same location, which may or may not have the same operator, and then generalise the results to a multiset of shared updates to the same location.

Definition 6.4.3. A binary operator μ_1 (over the domain D) is *compatible* to the binary operator μ_2 (notation: $\mu_1 \preceq \mu_2$) (over D) if

- μ_2 is associative and commutative, and
- for all $x \in D$, there is some $\dot{x} \in D$ such that

$$\text{for all } y \in D \text{ we have } y\,\mu_1\,x = y\,\mu_2\,\dot{x}.$$

Obviously, each associative and commutative operator μ is compatible to itself (simply we say *self-compatible*). The following Lemma gives a sufficient condition for compatibility.

Lemma 6.4.1. *Let μ_1 and μ_2 be two binary operators over domain D such that (D, μ_2) defines a commutative group, and $(x\,\mu_1\,y)\,\mu_2\,y = x$ holds for all $x, y \in D$. Then $\mu_1 \preceq \mu_2$ holds.*

Proof. Let $e \in D$ be the neutral element for μ_2, and let $\dot{x}$ be the inverse of x. Then we get

$$y\,\mu_1\,x = (y\,\mu_1\,x)\,\mu_2\,e = (y\,\mu_1\,x)\,\mu_2\,(x\,\mu_2\,\dot{x}) = ((y\,\mu_1\,x)\,\mu_2\,x)\,\mu_2\,\dot{x} = y\,\mu_2\,\dot{x}.$$

□

Example 6.4.3. Let us look back Example 6.4.1. Both $(\mathbb{R}, +)$ and $(\mathbb{Q}^*, \times)$ are abelian groups, the duality property in Lemma 6.4.1 is satisfied by addition $+$ and substraction $-$ on $\mathbb{R}$, and multiplication $\times$ and division $\div$ on $\mathbb{Q}^*$, respectively. Thus, $- \preceq +$ and $\div \preceq *$ hold on $\mathbb{R}$ and $\mathbb{Q}^*$, respectively.

Similarly, set operations such as union $\cup$, intersection $\cap$, symmetric difference $\diamond$ are self-compatible. Moreover, as $x - y = x \cap \bar{y}$ holds with the complement $\bar{y}$ of the set y, set difference $-$ is compatible to intersection $\cap$. □

Compatibility $\imath_1 \preceq \imath_2$ permits replacing each shared update $(\ell, v, \imath_1)$ by the shared update $(\ell, \dot{v}, \imath_2)$. Then the associativity and commutativity of $\imath_2$ guarantees order-independence. Thus, we obtain the following theorem.

Theorem 6.4.1. *A non-empty multiset $\ddot{\Delta}_\ell$ of shared updates on the same location ℓ is operator-compatible if*

- *either $|\ddot{\Delta}_\ell| = 1$ holds,*
- *or there exists a $\mu \in Opt(\ddot{\Delta}_\ell)$ such that,*

$$\text{for all } \mu_1 \in Opt(\ddot{\Delta}_\ell),\ \mu_1 \preceq \mu \text{ holds.}$$

Proof. The first case is trivial. In the second case, if $\mu_1 \preceq \mu$ holds, then we can replace all shared updates in $\ddot{\Delta}_\ell$ with μ_1 by shared updates with μ. In doing so, we obtain an update multiset, in which only the self-compatible operator μ is used. The associativity and commutativity of μ implies

$$(\dots((x\,\mu\,b_1)\,\mu\,b_2)\dots\mu\,b_k) = (\dots((x\,\mu\,b_{\varsigma(1)})\,\mu\,b_{\varsigma(2)})\dots\mu\,b_{\varsigma(k)})$$

for all $x, b_1, \dots, b_k$ and all permutations ς as desired.

□

Example 6.4.4. Suppose that we have $\ddot{\Delta}_{\ell_1}$ with $Opt(\ddot{\Delta}_{\ell_1}) = \{+,-\}$, $\ddot{\Delta}_{\ell_2}$ with $Opt(\ddot{\Delta}_{\ell_2}) = \{\times,\div\}$, $\ddot{\Delta}_{\ell_3}$ with $Opt(\ddot{\Delta}_{\ell_3}) = \{\cap,-\}$ and $\ddot{\Delta}_{\ell_4}$ with $Opt(\ddot{\Delta}_{\ell_4}) = \{\cap,\cup\}$.

From Theorem 6.4.1, we obtain that $\ddot{\Delta}_{\ell_1}$, $\ddot{\Delta}_{\ell_2}$ and $\ddot{\Delta}_{\ell_3}$ are operator-compatible, and $\ddot{\Delta}_{\ell_4}$ is not operator-compatible.

□

Remark 6.4.1. Theorem 6.4.1 allows checking the operator-compatibility of shared updates to the same location by only utilising the schema information. This approach can ensure conformance to the genericity principle of database transformations, while considerably improving database performance.

6.4.2 Normalisation Algorithm

By the notation $norm(\ddot{\Delta})$, we denote the normalisation of the update multiset $\ddot{\Delta}$. Furthermore, let Δ_λ be a *trivial update set*, indicating that an update set is inconsistent. This comes into play when we do not have operator-compatibility.

Normalisation of a given update multiset $\ddot{\Delta}$ is conducted for each location ℓ appearing in $\ddot{\Delta}$, i.e. we normalise $\ddot{\Delta}_\ell$. In doing so, $\ddot{\Delta}_\ell$ is transformed into a set containing exactly one exclusive update, provided $\ddot{\Delta}_\ell$ is operator-compatible. Otherwise $norm(\ddot{\Delta}_\ell) = \Delta_\lambda$. The following algorithm describes the normalisation process in detail.

Algorithm 6.4.1.
Input: An update multiset $\ddot{\Delta}$ and a state S
Output: An update set $norm(\ddot{\Delta})$
Procedure:

1. By scanning through updates in $\ddot{\Delta}$, the set of locations $Loc(\ddot{\Delta})$ appearing in $\ddot{\Delta}$ is obtained, shared updates to each location ℓ are put into $\ddot{\Delta}_\ell$ and all exclusive updates are put into an update set Δ_{excl}.

2. For each $\ddot{\Delta}_\ell$, the following steps are processed:

   (a) If $\ddot{\Delta}_\ell = \{\!\{(\ell, b, \mu)\}\!\}$, then $norm(\ddot{\Delta}_\ell) = \{(\ell, \mu(val_S(\ell), b)\}$;

   (b) otherwise, check $Opt(\ddot{\Delta}_\ell)$:

      i. If there exists $\mu' \in Opt(\ddot{\Delta}_\ell)$ such that for all $\mu \in Opt(\ddot{\Delta}_\ell)$, $\mu \preceq \mu'$ holds, then

         - translate each update $(\ell, b, \mu) \in \ddot{\Delta}_\ell$ where $\mu \neq \mu'$ into the form (ℓ, b', μ') according to the results from Lemma 6.4.1;
         - assume that the update multiset after finishing the translation on each update in $\ddot{\Delta}_\ell$ is $\{\!\{(\ell, b_1', \mu'), ..., (\ell, b_k', \mu')\}\!\}$, $\ddot{\Delta}_\ell$ can be integrated into the update set $norm(\ddot{\Delta}_\ell) = \{(\ell, b_\ell')\}$, where $b_\ell' = val_S(\ell)\ \mu'\ b_1'\ \mu'\ ...\ \mu'\ b_k'$

      ii. otherwise, $norm(\ddot{\Delta}) = \Delta_\lambda$ and then exit the algorithm.

3. $norm(\ddot{\Delta})$ is obtained by $norm(\ddot{\Delta}) = \bigcup_{\ell \in Loc(\ddot{\Delta})} norm(\ddot{\Delta}_\ell) \cup \ddot{\Delta}_{excl}$.

The following result is a direct consequence of the algorithm.

Corollary 6.4.2. *For an update multiset $\ddot{\Delta}$, its normalisation $norm(\ddot{\Delta})$ is different from Δ_λ iff $\ddot{\Delta}$ is operator-compatible.* □

If $norm(\ddot{\Delta}) = \Delta_\lambda$ for an update multiset $\ddot{\Delta}$, we can immediately draw the conclusion that $\ddot{\Delta}$ is not consistent. Otherwise, we obtain an update set containing only exclusive updates. In the following sections we will therefore assume $norm(\ddot{\Delta}) \neq \Delta_\lambda$, and investigate further inconsistencies among exclusive updates in an update set after normalisation.

6.5 Integration of Exclusive Updates

In this section, we will deal with the second stage of consistency checking, starting from a normalised update set that only contains exclusive updates. Since exclusive updates may have partial locations, the definition for the consistency of an update set can not be directly taken from the standard definition of ASMs. Even if, for values b and b' of any two exclusive updates to the same location ℓ in an update set Δ, we have $b = b'$, Δ still might not be consistent. It is possible that inconsistencies arise from updates of distinct but non-disjoint locations, as illustrated in Example 6.1.4. Therefore, instead of consistent, we call an update set value-compatible if such a condition is satisfied.

Definition 6.5.1. A set Δ of exclusive updates is *value-compatible* if, for each location ℓ in Δ, whenever $(\ell, b), (\ell, b') \in \Delta$ holds, we have $b = b'$.

An update set that contains exclusive updates may be value-compatible but not consistent. On the other hand, following the standard definition for the consistency of an update set, we can have the following fact.

Fact 6.5.1. Let $\Delta = \{(\ell_1, v_1), ..., (\ell_k, v_k)\}$ be an update set containing exclusive updates. If the condition $\bigwedge_{1 \leq i \neq j \leq k} \ell_i \not\sqsubseteq \ell_j$ is satisfied, then Δ is *consistent*.

Obviously, the condition $\bigwedge_{1 \leq i \neq j \leq k} \ell_i \not\sqsubseteq \ell_j$ is sufficient but not necessary. There are cases in which a set of exclusive updates to non-disjoint locations is consistent. Let us have a look at the following example.

Example 6.5.1. For the relation $I(R')$ in Example 6.1.3, suppose that we have

- $\Delta_1 = \{(f_{R'.A_2}(o_3)(0,1), b_{31}), (f_{R'.A_2}(o_3)(1,1), b_{32}), (f_{R'.A_2}(o_3), [b_{31}, b_{32}])\}$, where $(f_{R'.A_2}(o_3)(0,1), b_{31})$ adds b_{31} before the first element of the list $[b_3]$, $(f_{R'.A_2}(o_3)(1,1), b_{32})$ replaces the first element of the list $[b_3]$ and $(f_{R'.A_2}(o_3), (b_{31}, b_{32}))$ changes the list $[b_3]$ to $[b_{31}, b_{32}]$.

- $\Delta_2 = \{(f_{R'.A_3}(o_3)(c_{32}), 2), (f_{R'.A_3}(o_3)(c_{33}), 0), (f_{R'.A_3}(o_3), \{\!\{c_{31}, c_{32}, c_{32}\}\!\})\}$,

  where $(f_{R'.A_3}(o_3)(c_{32}), 2)$ changes the multiplicity of c_{32} in the multiset $\{\!\{c_{31}, c_{32}, c_{33}\}\!\}$ to be 2, $(f_{R'.A_3}(o_3)(c_{33}), 0)$ removes c_{33} from the multiset and $(f_{R'.A_3}(o_3), \{\!\{c_{31}, c_{32}, c_{32}\}\!\})$ changes the multiset $\{\!\{c_{31}, c_{32}, c_{33}\}\!\}$ to $\{\!\{c_{31}, c_{32}, c_{32}\}\!\}$.

In this example, both Δ_1 and Δ_2 are consistent, although they both contain non-disjoint locations. In Δ_1, applying the updates $(f_{R'.A_2}(o_3)(0,1), b_{31})$ and $(f_{R'.A_2}(o_3)(1,1), b_{32})$ simultaneously over the relation $I(R')$ leads to the update $(f_{R'.A_2}(o_3), [b_{31}, b_{32}])$, which coincides with the third update in Δ_1. Similarly, applying the updates $\{(f_{R'.A_3}(o_3)(c_{32}), 2)$ and $(f_{R'.A_3}(o_3)(c_{33}), 0)$ over the relation $I(R')$ leads to an update, which coincides with the third update $(f_{R'.A_3}(o_3), \{\!\{c_{31}, c_{32}, c_{32}\}\!\})$ in Δ_2. Therefore, Δ_1 and Δ_2 are both consistent. □

The above example demonstrates that, in order to check the consistency of exclusive updates that may have non-disjoint locations, exclusive updates which have locations as the sublocations of another location at one-level higher abstraction should be integrated together. To develop a systematic solution for checking the consistency of a set of exclusive updates, we start by the parallel composition operations for partial updates to locations constructed by using common type constructors set, multiset, list and tuple.

6.5.1 Parallel Composition

The purpose of defining parallel composition operations is to capture the observed compositional relationship among sublocations established by the use of type constructors in a data model. The significance of type constructors in the parallel composition operation for an update set can be illustrated by the following example.

Example 6.5.2. The sublocations of the prime location $R'(\{(a_{31}, a_{32})\}, [b_3], \{\!\{c_{31}, c_{32}, c_{33}\}\!\})$ presented in Example 6.3.3 are constructed by applying various type constructors, which are illustrated in Figure 6.6. The notations $\{\cdot\}$, $\{\!\{\cdot\}\!\}$, $[\cdot]$ and $(\cdot)^n$ denote the type constructors for sets, multisets, lists and n-ary tupes, respectively.

For instance, the locations ℓ_{111} and ℓ_{112} refer to the values of attributes A_{11} and A_{12} of the subtuple represented by the location ℓ_{11}, respectively. They are constructed into the location ℓ_{11} by applying a 2-ary tuple constructor.

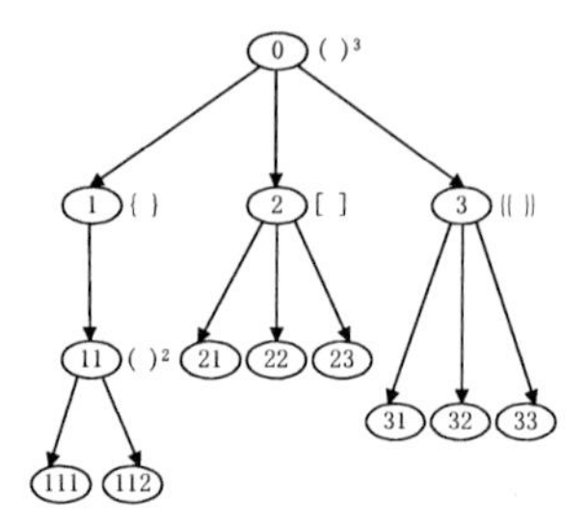

Figure 6.6: Constructors in partial locations

□

Now we define the parallel composition operations for partial updates on a case-basis by considering each type constructor for which locations has been defined. The focus is on the main type constructors used in a data model of complex-value databases which are set, multiset, list and tuple.

Set

Set constructor has been widely used in various data modeling. Assume that we have a location ℓ representing a set in a state S, i.e., $val_S(\ell) = f$, and locations referring to the elements $\overline{a}$ of the set f are expressed as $f(\overline{a})$. For the set Δ of updates in which the locations refer only to the elements of the set f, if Δ is value-compatible, then the set of updates in Δ can be integrated into an update (ℓ, b) such that

$$\Omega(\Delta) = (\ell, b),$$

where

$$\begin{aligned} b = val_S(\ell) \cup \{\overline{a}_i | b_i = true \wedge (f(\overline{a}_i), b_i) \in \Delta\} \\ -\{\overline{a}_i | b_i = false \wedge (f(\overline{a}_i), b_i) \in \Delta\}. \end{aligned}$$

Multiset

Multiset constructor is also known as bag constructor in data modeling. Assume that we have a location ℓ representing a multiset in a state S, i.e., $val_S(\ell) = \mathcal{M}$. As discussed in

Section 6.1, a multiset $\mathcal{M}$ may be represented as a set of elements of the form (a, c) where a is the element of $\mathcal{M}$ and c its multiplicity in the multiset $\mathcal{M}$, and a location referring to an element a of the multiset $\mathcal{M}$ is expressed as $f_{\mathcal{M}}(a)$. A value-compatible set Δ of updates in which the locations refer only to the elements of the multiset $\mathcal{M}$ can be integrated into an update (ℓ, b) such that

$$\Omega(\Delta) = (\ell, b),$$

where

$$b = val_S(\ell) - \{(a, b')|(a, b') \in val_S(\ell) \wedge (f_{\mathcal{M}}(a), b) \in \Delta \wedge b \neq b'\}$$
$$\cup\{(a, b)|(f_{\mathcal{M}}(a), b) \in \Delta\}.$$

List

List constructor provides the capability of modelling the order of elements when such an order is of interest. Consequently, the sublocations constructed by applying a list constructor are ordered, which we can capture by using a strict dependence relation $\trianglelefteq$ among them as discussed in Section 6.3.1. Assume that we have a location ℓ representing a list s in a state S, and the locations referring to the parts of the list are expressed by $f_s(k_1, k_2)$ as discussed in Section 6.1. Then, a value-compatible set $\Delta = \{(\ell_1, b_1), ..., (\ell_n, b_n)\}$ of updates, in which the locations ℓ_i $(i = 1, ..., n)$ refer only to the elements of the list s can be integrated into an update (ℓ, b) such that

$$\Omega(\Delta) = (\ell, b),$$

where

$$b = val_{S+\{(\ell_{p_1}, b_{p_1})\}+...+\{(\ell_{p_n}, b_{p_n})\}}(\ell) \text{ and } \ell_{p_i} \trianglelefteq \ell_{p_{i+1}}$$

for a permutation $p_1, \ldots, p_n$ of the updates in Δ and $i = 1, \ldots, n-1$. That is, b is the list obtained by applying Δ over the list s in the current state S in the order of first taking the update whose location is being dependent by the locations of other updates.

Tuple

Tuple constructor can be treated in a similar way to list constructor, except that the order of applying updates in an update set can be arbitrarily chosen. Assume that the location ℓ representing a tuple in a state S. Then, a value-compatible set $\Delta = \{(\ell_1, b_1), ..., (\ell_n, b_n)\}$ of updates, in which the locations refer only to the attribute values of the tuple represented by ℓ can be integrated into an update (ℓ, b) such that

$$\Omega(\Delta) = (\ell, b),$$

where

$$b = val_{S+\Delta}(\ell).$$

6.5.2 Location-Based Partitions

In order to efficiently handle dependencies between partial locations, we propose to partition a given update set containing only exclusive updates into a family of update sets. Each update set in such a family is called a cluster which has an update subsuming all other updates. The formalisation is provided as follows.

The notation $SubL(\ell)$ denotes the set of all the sublocations of a location ℓ.

Lemma 6.5.1. *Let $\mathcal{L}_S$ denote the set of locations in a state S. Then there exists a unique partition $\mathcal{L}_S = \bigcup_{i \in \mathrm{I}} \mathcal{L}_i$ such that*

- *for all $i, j \in \mathrm{I}$ with $i \neq j$ we have $\ell_i \not\trianglelefteq \ell_j$ and $\ell_j \not\trianglelefteq \ell_i$ for all $\ell_i \in \mathcal{L}_i$ and $\ell_j \in \mathcal{L}_j$, and*
- *for each $i \in \mathrm{I}$ there exists a location $\ell_i \in \mathcal{L}_S$ with $\mathcal{L}_i = SubL(\ell_i)$.*

Proof. By taking connected components of the graph defined by $(\mathcal{L}_S, \trianglelefteq)$, we can partition $\mathcal{L}_S$ into $\mathcal{L}_i$ ($i \in \mathrm{I}$) satisfying the first property. Moreover, none of the $\mathcal{L}_i$ can be further decomposed while still satisfying the first property and we cannot combine multiple partition classes such that the second property holds. Thus, this partition is unique.

According to the definition of the subsumption relation $\sqsubseteq$, each $SubL(\ell)$ is contained in one $\mathcal{L}_i$, and $SubL(\ell_2) \subseteq SubL(\ell_1)$ holds for $\ell_2 \sqsubseteq \ell_1$. On the other hand, maximal elements with respect to $\sqsubseteq$ define disjoint locations. Therefore, for a maximal element ℓ with respect to $\sqsubseteq$ we must have $SubL(\ell) = \mathcal{L}_i$ for some $i \in \mathrm{I}$, which shows the second property.

□

Now let Δ be an update set containing exclusive updates. Using the partition of $\mathcal{L}_S$ from Lemma 6.5.1, we obtain a partition $\Delta = \bigcup_{i \in I'} \Delta_i$ where $\Delta_i = \{(\ell, b) \in \Delta \mid \ell \in \mathcal{L}_i\}$ and $I' = \{i \in I \mid \Delta_i \neq \emptyset\}$. The following lemma is a direct consequence of the independence of locations in different set $\mathcal{L}_i$.

Lemma 6.5.2. *Δ is consistent iff each Δ_i for $i \in I'$ is consistent.*

As not all locations in $\mathcal{L}_i$ appear in an update set Δ, we may further decompose each Δ_i for $i \in I'$. For this, let $\mathcal{L}(\Delta_i) \subseteq \mathcal{L}_i$ be the set of locations appearing in Δ_i. By taking connected components of the graph defined by $(\mathcal{L}(\Delta_i), \sqsubseteq)$ we can get partition $\mathcal{L}(\Delta_i) = \bigcup_{j \in J_i} \mathcal{L}_{ij}$ such that for all $j_1, j_2 \in J_i$ with $j_1 \neq j_2$ we have $\ell_{j_1} \not\sqsubseteq \ell_{j_2}$ and $\ell_{j_2} \not\sqsubseteq \ell_{j_1}$ for all $\ell_{j_1} \in \mathcal{L}_{ij_1}$ and $\ell_{j_2} \in \mathcal{L}_{ij_2}$. As none of the $\mathcal{L}_{ij}$ can be further decomposed, this partition is also unique. Taking $\Delta_{ij} = \{(\ell, b) \in \Delta_i \mid \ell \in \mathcal{L}_{ij}\}$ and omitting those of these update sets that are empty, we obtain a unique partition of Δ_i.

Lemma 6.5.3. *Δ_i is consistent for $i \in I'$ iff each Δ_{ij} with $j \in J_i$ is consistent.*

Proof. Consider the maximal elements $\ell_{i1}, ..., \ell_{ik}$ in $\mathcal{L}(\Delta_i)$ with respect to $\sqsubseteq$ and the unique values v_{ij} ($j = 1, ..., k$) with $(\ell_{ij}, v_{ij}) \in \Delta_i$. Let S be a state with $val_S(\ell_i) = v_i$. If Δ_{ij} is consistent, then $val_{S+\{(\ell_{ij}, v_{ij})\}}(\ell) = val_{S+\Delta_{ij}}(\ell)$ for all $(\ell, v) \in \Delta_{ij}$. As the locations ℓ_{ij} are pairwise disjoint, according to Fact 6.5.1 we may simultaneously apply all updates (ℓ_{ij}, v_{ij}) to v_i to obtain a value v_i', thus $val_{S+\{(\ell_i, v_i')\}}(\ell) = val_{S+\{(\ell_{i1}, v_{i1}), ..., (\ell_{ik}, v_{ik})\}}(\ell)$ for all $(\ell, v) \in \Delta_i$.

The converse, that Δ_i (i.e., the union of all Δ_{ij}) is not consistent if any Δ_{ij} is not consistent, is obvious. □

In the proof, we actually showed more, as we only need "upward consistency" for the set of locations below the maximal elements ℓ_{ij}.

Corollary 6.5.1. *For the maximal elements $\ell_{i1}, \ldots, \ell_{ik}$ in $\mathcal{L}(\Delta_i)$ with respect to $\sqsubseteq$, let $\Delta_{ij} = \{(\ell, v) \in \Delta_i \mid \ell \sqsubseteq \ell_{ij}\}$. Then Δ_i is consistent iff all Δ_{ij} ($j = 1, \ldots, k$) are consistent.* □

Note that the update sets Δ_{ij} in Corollary 6.5.1 are uniquely determined by Δ. There exist locations ℓ_i and ℓ_{ij} such that $\ell_{ij} \sqsubseteq \ell_i$ and for all updates $(\ell, v) \in \Delta_{ij}$ we have $\ell \sqsubseteq \ell_{ij}$. We call such an update set Δ_{ij} a *cluster* below ℓ_{ij}.

With respect to the subsumption relation $\sqsubseteq$, locations in $\mathcal{L}_i$ may be assigned with levels. Assume that the length of the longest downward path to a minimal element from the maximal element in $\mathcal{L}_i$ is n. Then,

- the maximal element is a location at the level n,
- the elements which are the children of a location at the level k are locations at the level $k-1$.

Thus, the maximal element $\ell_i \in \mathcal{L}_i$ (as in Lemma 6.5.1) resides at the highest level, the minimal element in $\mathcal{L}_i$ resides at the lowest level and other locations in $\mathcal{L}_i$ are arranged at levels in the middle. A location ℓ at the level n is denoted as ℓ^n. For a cluster Δ_{ij} below ℓ_{ij}, the level of ℓ_{ij} is called the *height* of Δ_{ij} and is denoted as $height(\Delta_{ij})$.

Example 6.5.3. Let us consider again the prime location $R'(\{(a_{31}, a_{32})\}, [b_3], \{\!\{c_{31}, c_{32}, c_{33}\}\!\})$ and its sublocations (see Example 6.3.3). The pictures (a)-(c) of Figure 6.7 illustrate three scenarios as follows:

- In picture (a), suppose that we have $\Delta = \{(\ell_{112}, a'_{32}), (\ell_{22}, b_{31}), (\ell_{23}, b_{32})\}$. Because ℓ_{112}, ℓ_{22} and ℓ_{23} do not subsume one another, so Δ is partitioned into three clusters: $\{(\ell_{112}, a'_{32})\}$ below ℓ_{112}, $\{(\ell_{22}, b_{31})\}$ below ℓ_{22} and $\{(\ell_{23}, b_{32})\}$ below ℓ_{23} as circled out in the picture. Note that, although $\ell_{22} \trianglelefteq \ell_{23}$, they are disjoint and thus consistent in this case.
- In picture (b), suppose that we have $\Delta = \{(\ell_{112}, a'_{32}), (\ell_{22}, b_{31}), (\ell_{23}, b_{32}), (\ell_2, [b_{31}, b_{32}])\}$. Because $\ell_{22} \sqsubseteq \ell_2$ and $\ell_{23} \sqsubseteq \ell_2$, thus (ℓ_{22}, b_{31}), (ℓ_{23}, b_{32}) and $(\ell_2, [b_{31}, b_{32}])$ are partitioned into one cluster, while (ℓ_{112}, a'_{32}) is in another cluster by itself. In this way, we can check the consistency of the cluster below ℓ_2 by checking the consistency of the result of integrating (ℓ_{22}, b_{31}) and (ℓ_{23}, b_{32}) with $(\ell_2, [b_{31}, b_{32}])$.
- In picture (c), we consider the update set $\Delta = \{(\ell_{112}, a'_{32}), (\ell_{22}, b_{31}), (\ell_{23}, b_{32}), (\ell_2, [b_{31}, b_{32}]), (\ell_0, (\emptyset, [b_3], \{\!\{c_{31}, c_{32}\}\!\}))\}$. As ℓ_{112}, ℓ_{22}, ℓ_{23} and ℓ_2 are all subsumed by the location ℓ_0, they are all in one cluster.

□

6.5.3 Cluster-Compatibility

In light of Corollary 6.5.1, the problem of consistency checking is reduced to that of verifying the consistency of clusters.

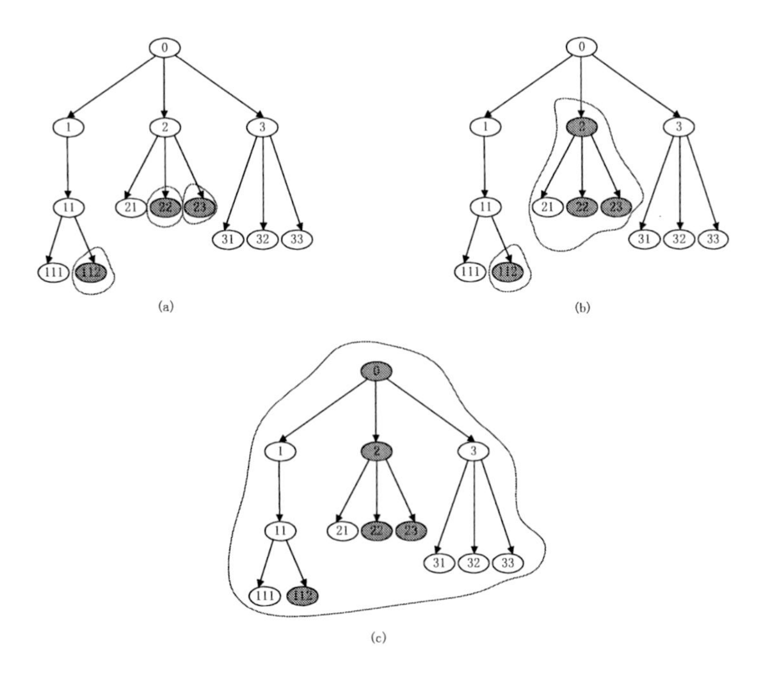

Figure 6.7: Location-based partitions and clusters

Lemma 6.5.4. *Let Δ^ℓ be a cluster below the location ℓ. If the set $\{(\ell_1^n, b_1), ..., (\ell_i^n, b_i)\}$ of all updates in Δ^ℓ at a level $n < height(\Delta^\ell)$ is value-compatible, then, as discussed in Subsection 6.5.1, it is possible to define a set $\{(\ell_1^{n+1}, b_1'), ..., (\ell_j^{n+1}, b_j')\}$ of updates at the level $n+1$ such that, for all states S and any location $\ell' \in \mathcal{L}_S$, we have*

$$val_{S+\{(\ell_1^n,b_1),...,(\ell_i^n,b_i)\}}(\ell') = val_{S+\{(\ell_1^{n+1},b_1'),...,(\ell_j^{n+1},b_j')\}}(\ell').$$

Proof. Since the level n is less than $height(\Delta^\ell)$, the set $\{(\ell_1^n, b_1), ..., (\ell_i^n, b_i)\}$ of updates can be grouped based on the condition whether their locations are subsumed by the same location at the level $n+1$, e.g., $\{(\ell_{k_1}^n, b_{k_1}), ..., (\ell_{k_p}^n, b_{k_p})\} \subseteq \{(\ell_1^n, b_1), ..., (\ell_i^n, b_i)\}$ is the group in which the locations $\ell_{k_1}^n, ..., \ell_{k_p}^n$ are subsumed by some location $\ell_m^{n+1} \in \{\ell_1^{n+1}, \ldots, \ell_j^{n+1}\}$. Then, for each group of updates to the locations at the level n, if they are value-compatible, then they can be integrated into an exclusive update to a location at the level $n+1$ as follows:

$$\Omega\{(\ell_{k_1}^n, b_{k_1}), ..., (\ell_{k_p}^n, b_{k_p})\} = (\ell_m^{n+1}, b_m'),$$

where $val_{S+\{(\ell_{k_1}^n,b_{k_1}),...,(\ell_{k_p}^n,b_{k_p})\}}(\ell') = val_{S+\{(\ell_m^{n+1},b_m')\}}(\ell')$ for each state S and all $\ell' \in \mathcal{L}_S$.

In doing so, the set of updates $\{(\ell_1^n, b_1), ..., (\ell_i^n, b_i)\}$ defines a set of exclusive updates $\{(\ell_1^{n+1}, b_1'), ..., (\ell_j^{n+1}, b_j')\}$ in which the locations are one level higher than n if it is value-compatible.

□

We finally obtain the following main result on the consistency of clusters.

Theorem 6.5.1. *Let Δ^ℓ be a cluster below the location ℓ. If Δ^ℓ is "level-by-level" value-compatible, then Δ^ℓ is consistent.*

Proof. If Δ^ℓ is "level-by-level" value-compatible, then for any state S and starting from updates on locations at the lowest level, exclusive updates to locations at the same level in Δ^ℓ can be replaced by exclusive updates to one-level-higher locations as stated in Lemma 6.5.4. As the set of exclusive updates at each level is value-compatible, this procedure continues until we reach the highest level in Δ^ℓ, i.e., the height of Δ^ℓ. Finally, all the updates at the level Δ^ℓ are combined into a single exclusive update (ℓ, b) if they are value-compatible, i.e., $val_{S+\{(\ell,b)\}}(\ell') = val_{S+\Delta}(\ell')$ for all $\ell' \in \mathcal{L}_S$.

□

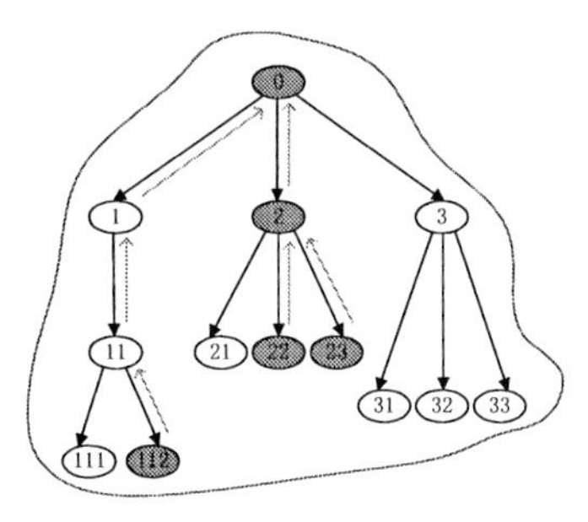

Figure 6.8: The consistency checking of a cluster

Example 6.5.4. Let us look back again the cluster below the location ℓ_0 in picture (c) of Figure 6.7. First, $\{(\ell_{112}, a'_{32})$ at level 0 can be integrated into update $(\ell_{11}, (a_{31}, a'_{32}))$ at level 1. Then $(\ell_{11}, (a_{31}, a'_{32}))$ at level 1 is integrated into update $(\ell_1, \{(a_{31}, a'_{32})\})$ at level 2, and similarly, integrating (ℓ_{22}, b_{31}) and (ℓ_{23}, b_{32}) at level 2 results in update $(\ell_2, [b_{31}, b_{32}])$ at level 2, which is identical with the original update to the location ℓ_2 in the cluster. As $(\ell_1, \{(a_{31}, a'_{32})\})$ and $(\ell_2, [b_{31}, b_{32}])$ are also value-compatible, they can be integrated to check for consistency against $(\ell_0, (\emptyset, [b_3], \{\!\{c_{31}, c_{32}\}\!\}))$. Since the resulting update $(\ell_0, (\{(a_{31}, a'_{32})\}, [b_{31}, b_{32}], \{\!\{c_{31}, c_{32}, c_{33}\}\!\}))$ at level 3 is not value-compatible with the update $(\ell_0, (\emptyset, [b_3], \{\!\{c_{31}, c_{32}\}\!\}))$ at level 3, thus this cluster above ℓ_0 is not consistent.

In Figure 6.8, several arrows are added to illustrate the above procedure.

□

6.5.4 Integration Algorithm

In this subsection, we present how to algorithmically integrate exclusive updates. For clarity, the procedure is given in terms of two algorithms:

- The first algorithm clusters the updates in a given set of exclusive updates. Every update is initially assumed to define a cluster. We then successively consider each pair of updates where one update subsumed the other, and amalgamate their respective clusters into larger ones until no more changes can be made.

Algorithm 6.5.1.

Input: An update set Δ that only contain exclusive updates

Output: A set $clus(\Delta)$ of clusters

Procedure:

1. starting with $P = \emptyset$ and $clus(\Delta) = \{\{u\}|u \in \Delta\}$;
2. checking the subsumption relation for any two updates $u_x, u_y \in \Delta$,
   - if the locations of u_x and u_y are related by subsumption, then add $\{u_x, u_y\}$ into P such that $P = P \cup \{\{u_x, u_y\}\}$;
3. doing the following as long as there are changes to $clus(\Delta)$:
   - for each element V in P, do the following
     * $V' = \bigcup\{x|x \in clus(\Delta)$ and $x \cap V \neq \emptyset\}$,
     * $clus(\Delta)$=$clus(\Delta) \cup \{V'\} - \{x|x \subseteq V'$ and $x \in clus(\Delta)\}$.

- The second algorithm then take the set of clusters and transforms it into a set of exclusive updates in which locations are pairwise disjoint. This is done in accordance with Theorem 6.5.1, that is, through level-by-level integration provided the updates in each cluster at each level is value-compatible.

Algorithm 6.5.2.

Input: A set $clus(\Delta)$ of clusters

Output: An update set Δ

Procedure:

1. $\Delta = \emptyset$;
2. For each cluster $\Delta_i \in clus(\Delta)$, apply the following steps:
   - Assigning a level to each location in $Loc(\Delta_i)$ in accordance with the schema information provided by the database environment;
   - $V = \Delta_i$;
   - Doing the following until the height of the cluster Δ_i is reached:
     * $P = \{(\ell, b)|(\ell, b) \in V$ and the level $level(\ell)$ of ℓ is minimal in $V\}$;

  * partition updates in P such that, for each partition class $\{(\ell_1, b_1), ..., (\ell_n, b_n)\} \subseteq P$, there exists a location ℓ with $level(\ell) = \ell_i + 1$ and $\ell_i \sqsubseteq \ell$ $(i = 1, ..., n)$;
  * For each partition class $\{(\ell_1, b_1), ..., (\ell_n, b_n)\} \subseteq P$, checking the value-compatibility of the update set $\{(\ell_1, b_1), ..., (\ell_n, b_n)\}$.
    (a) if it is value-compatible, then do the following:
      · apply the parallel composition operation $(\ell, b) = \Omega\{(\ell_1, b_1), ..., (\ell_n, b_n)\}$;
      · $V = V - P \cup \{(\ell, b)\}$.
    (b) otherwise, $\Delta = \Delta_\lambda$ and then exit the algorithm.
- $\Delta = \Delta \cup V$.

3. Exit the algorithm with Δ.

The whole algorithm for the integration of exclusive updates, denoted as Algorithm$_{Inte}$, can then be considered as the function composition of the above two algorithms such that

$$\text{Algorithm}_{Inte}\text{=Algorithm 6.5.2} \circ \text{Algorithm 6.5.1}$$

6.6 Applications in Aggregate Computing

Aggregate computing acquires a central role in many distributed database applications. The result of applying an aggregate function is calculated from a collection of elements in a state. Quite often, such a result can be partially computed at some point and then combined together later. This phenomenon poses the question: what role partial updates might play in aggregate computing. In this section, we will exploit partial updates in optimising, rewriting and maintaining aggregate computations.

We begin with a general formalism for aggregate functions as defined by Cohen [60], which covers both built-in and user-defined aggregate functions appearing in database applications. As stated in Chapter 3, a location operator of DB-ASMs is defined to be such a generalised aggregate function. That is, the assignment of an aggregate function $\rho(m)$ to a term t, in the form of $t := \rho(m)$ for a multiset $m = \{\!\{x|\varphi(x)\}\!\}$, can be equivalently expressed by the DB-ASM rule as shown at the left hand side of Figure 6.9. The DB-ASM rule first generates a multiset of updates such as $\{\!\{(t, x)|\varphi(x)\}\!\}$, then the update multiset is collapsed to an update set containing a single update $(t, \rho(m))$ by applying the location

let $\theta(t) = \rho$
  forall x with $\varphi(x)$
  do
    $t := x$
  enddo
endlet
	forall x with $\varphi(x)$
	do
	  $t' \Leftarrow^{\odot} f_{\beta_1}(x)$
	enddo;
	$t \Leftarrow f_{\beta_2}(t')$

Figure 6.9: Aggregate functions, location operators and partial updates

operator ρ as specified in Definition 3.2.1. Therefore, we can simply treat a location operator as an aggregate function in our discussion.

An alternative and more natural view of aggregate computing involves partial updates. Suppose that $\rho = (f_{\beta_1}, \odot, f_{\beta_2})$. Then the DB-ASM rule at the left hand side of Figure 6.9 can be expressed by using the partial update rules at the right hand side. Of note is that, the partial update rules at the right hand side generate partial updates, whereas the DB-ASM rule at the left hand side generates standard updates. Following Theorem 6.4.1, all shared updates generated in the partial update rules at the right hand side of Figure 6.9 is operator-compatible because of the associativity and commutativity stipulated on $\odot$ in Definition 3.1.9. This fact manifests that partial updates are the essential part of aggregate operations and also illustrates that the partial update rules can be encoded as DB-ASMs rules. Thus, partial updates become less noteworthy in the setting of DB-ASMs[1]. However, the existence of the partial update rules is important for providing natural, flexible and practical database transformations.

The properties of aggregation functions and their correspondence to database applications have been extensively studied in the literature, e.g., [60, 158, 118]. Amongst these properties, the class of decomposable aggregation functions is of most interest to us due to its connection with partial updates. Let $\rho = (f_{\beta_1}, \odot, f_{\beta_2})$ be an aggregate function and $m = m_1 \uplus \cdots \uplus m_n$ be a multiset m with its submultisets $m_1, \ldots, m_n$. Then ρ is called *decomposable* if the following condition holds:

$$\rho(m) = \rho(m_1) \odot \cdots \odot \rho(m_n).$$

Most aggregate functions used in database applications are decomposable.

[1]As noted in Remark 10.4. of the paper [27] characterising parallel algorithms, partial updates can be eliminated in a way by viewing them as messages sent to proclets and then integrating all relevant partial updates into appropriate total updates, where locations are disjoint as defined in a standard way.

par
  forall x in $\varphi_1(x)$ do
    $t_1^{'} \leftleftarrows^{\odot \circ f_{\beta_2}} f_{\beta_1}(x)$
  enddo
  forall x in $\varphi_2(x)$ do
    $t_2^{'} \leftleftarrows^{\odot \circ f_{\beta_2}} f_{\beta_1}(x)$
  enddo
par;
$t \leftleftarrows t_1^{'} \odot t_2^{'}$

Figure 6.10: Decomposable aggregate functions and partial updates

Example 6.6.1. Let f_{id} be an identity function, f_{card} and $f_{\geq}$ be functions such that $f_{card}(x) = 1$ and $f_{\geq}(x, y) = x$ if $x \geq y$, otherwise $f_{\geq}(x, y) = y$. The aggregation functions *sum*, *count* and *max* are decomposable, such that, for $m = m_1 \uplus m_2$, we have

- for $sum = (f_{id}, +, f_{id})$, $sum(m) = sum(m_1) + sum(m_2)$,
- for $count = (f_{card}, +, f_{id})$, $count(m) = count(m_1) + count(m_2)$, and
- for $max = (f_{id}, f_{\geq}, f_{id})$, $max(m) = max(m_1)\ f_{\geq}\ max(m_2)$.

□

It is straightforward to prove that ρ is decomposable iff f_{β_2} is distributive over $\odot$, i.e., $f_{\beta_2}(x_1 \odot x_2) = f_{\beta_2}(x_1) \odot f_{\beta_2}(x_2)$. Hence, an aggregation function in the form of $(f_{\beta_1}, \odot, f_{id})$ must be decomposable. When $f_{\beta_2} \neq f_{id}$ but it is distributive over $\odot$, an aggregate function $\rho = (f_{\beta_1}, \odot, f_{\beta_2})$ can be transformed into an equivalent form $\rho^{'} = (f_{\beta_1}, \odot \circ f_{\beta_2}, f_{id})$. Figure 6.10 presents a program with the partial update rules, corresponding to the decomposition of aggregate functions.

The decomposable aggregation functions are significant for query rewriting and view maintenance in aggregate computing, as they possess some nice features. First of all, it enables an aggregate operation over a multiset m to be split into multiple aggregate operations over submultisets of m, executing in parallel or individually. In doing so, we can optimise performance with parallelism, implement incremental maintenance for views and rewrite queries based on materialised views. Secondly, an aggregate function can be applied to a large multiset of elements recursively, i.e. a submultiset or a submultiset of a

forall x_4 **with** $\exists x_1, x_2, x_3$. PERSON(x_1, x_2, x_3, x_4)
do
  forall x_2' **with** $\exists x_1', x_3'$. PERSON(x_1', x_2', x_3', x_4)
  do
    $y \Leftarrow^{|,} x_2'$
  enddo;
  result$(x_4, y) \Leftarrow$ *true*
enddo

Figure 6.11: Non-deterministic aggregate function *concat*()

submultiset of elements at each time. This can provide a flexible mechanism to maximise the parallelism of a computation, customised for a specific environment. In both cases, partial computations via a splitable collection of partial updates is a key requirement.

Note that, although non-decomposable aggregation functions cannot help with the rewrite of queries or incremental maintenance of views, they may still be used to improve performance to some degree by first computing the partial results in parallel with f_{β_1} and $\odot$, and processing the combined results with f_{β_2} at some point later.

The requirement that $\odot$ of an aggregate function be commutative and associative only applies for deterministic aggregate functions. When non-determinism is permitted, this requirement turns out to be too stringent. For instance, similar to the way the aggregate function *sum* defined over real numbers, a concatenation of elements from a multiset is often needed by database users, which is defined in the form of a non-deterministic aggregate function $concat = (f_{id}, |_\star, f_{id})$, where $|_\star$ is a non-commutative but associative function such that $|_\star(x, y) = x \star y$. Clearly, different orderings of applying $|_\star$ over elements give rise to different results.

Example 6.6.2. Let us consider the following relation I(PERSON) over PERSON = {PersonID,Fname,Sname,Address} in Figure 6.12, and the query

"Show the first names of people who have the same address".

In the rules of Figure 6.11, the aggregate function *concat*() is used to concatenate a set of first names together. Due to non-determinism, there are several possible results. Two of them are presented in Figure 6.12.

□

- I(PERSON):

PersonID	Fname	Sname	Address
1	Tom	Adler	70 King Street
2	Jack	Baird	16 Blacks Road
3	Jia	Adler	70 King Street
4	Mike	Cone	16 Blacks Road
5	Peter	Cone	16 Blacks Road

- Two of possible results:

Address	Fname
70 King Street	Tom,Jia
16 Blacks Road	Jack,Mike,Peter

Address	Fname
70 King Street	Jia,Tom
16 Blacks Road	Mike,Peter,Jack

Figure 6.12: A relation I(PERSON) and two possible query results

A possible solution for capturing this kind of non-deterministic aggregate function in database transformations would be to generalise the formalism $(f_{\beta_1}, \odot, f_{\beta_2})$ of aggregate functions by removing the commutative property defined on $\odot$. Consequently, the process of normalising shared updates needs to be revised, which we leave for future work.

Chapter 7

Conclusion

This chapter summarises the main results contained in the dissertation and reflects on the current work to suggest several interesting directions for future research.

7.1 Summary of Main Results

This dissertation presents a foundational study of database transformations over complex-value databases.

Instead of defining database transformation as computable functions in the recursion-theoretic sense, we aim to characterise them as algorithms. Through the investigation, a unifying theoretical framework for database transformations that encompass both queries and updates has been established by using the theory of Abstract State Machines. The development of this unifying framework is significant, as it bridges the gap between the well-established theoretical foundations for database transformations and various practical toolkits for database applications.

We presented a variant of Gurevich's sequential ASM thesis [82] dealing with database transformations in complex-value databases. Analogous to Gurevich's seminal work, we formulated several intuitive postulates for database transformations and discussed why database transformations should satisfy these postulates. Ultimately, five postulates were defined: the sequential time postulate, the abstract state postulate, the background postulate, the bounded exploration postulate and the genericity postulate. We then defined a variant of Abstract State Machines, which is called database Abstract State Machines (DB-ASMs). It has been showed that DB-ASMs capture exactly all database transformations characterised by the five postulates.

Despite many little technical differences of minor importance – such as final states, finite runs, and undefined successor state in case of an inconsistent update set – we stayed rather close to Gurevich's seminal work. Any extension reflects what we felt is necessary to capture the essential features of database transformations as opposed to sequential algorithms. Some important differences to Gurevich's sequential ASM thesis are:

- the permission of non-determinism in a limited form in which the limitations are enforced by the genericity postulate,
- the exploitation of meta-finite states to capture the finiteness of databases along with a possibly infinite algorithmic structure for facilitating computations by the abstract state postulate,
- the use of background classes to capture specific data model requirements by the background postulate,
- the presence of non-ground terms in a bounded exploration witness by an extended bounded exploration postulate, and
- the incorporation of location operators with let rules to support aggregate computing.

Through refinements, DB-ASMs can be used as the basis for studying database transformations in different data models. An important step in being able to do so is to make the necessary background for a data model explicit. We have made the backgrounds explicit for the cases of relational and XML database transformations in Chapter 4. For relational database transformations, it turns out that finite set, tuple and multiset constructors must be included in backgrounds to support database transformations in relational models. Furthermore, operators of relational algebra must exist in the backgrounds of relational database transformations which utilise relational algebra to optimise the implementation of relational database transformations based on a set of algebraic identities. For XML database transformations, we adopted hereditarily finite trees, operators of a hedge algebra, weak monadic second-order logic and extended document type definitions in tree-based backgrounds. Consequently, DB-ASMs with backgrounds defined in these particular ways capture relational and XML database transformations, respectively. In doing so, we actually proved that capturing specific characteristics of data models by background classes is the appropriate way to develop the theory, this is in contrast to our earlier work [154] where we tried to approach the problem by direct manipulation of the notion of state to deal with higher-order structures, instead of first-order ones.

We did, however, go one step further [153, 156]. An alternative and more elegant computational model for XML database transformations is defined, which directly incorporates weak MSO formulae in forall and choice rules. This leads to so-called XML machines. Due to the intuition behind the postulates, it should come as no surprise that XML machines and DB-ASMs with tree-based backgrounds have in fact equivalent expressive power. Their equivalence was proven by showing that XML machines satisfy the postulates of database transformations with tree-based backgrounds.

In order to investigate the characterisation of database transformations in a precise and mathematical way, we proposed a logic for DB-ASMs. First steps in this direction have been made in [155]. Built upon a logic of meta-finite states, the logic for DB-ASMs was formalised with atomic predicates for the definedness of a DB-ASM rule, for updates in an update set and multiset, and for an update set and multiset yielded by DB-ASM rules, modal operators for an update set and second-order quantifiers. Although the logic for DB-ASMs is developed in a similar way to the logic for ASMs [134], the logic for DB-ASMs solved the problem in the formalisation of consistency for non-deterministic transitions, which was encountered in the logic for ASMs. This is due to the finiteness condition stipulated in the abstract state postulate and the use of modal operators for update sets, instead of modal operators for rules as used in ASMs. Furthermore, by parameterising the logic of meta-finite states with the first-order logic, we obtained a sound and complete proof system for the logic for DB-ASMs. It allows us to reason about database transformations specified by DB-ASMs.

Finally, we addressed the problem of partial updates in the context of complex-value databases. The work was motivated by the need for an efficient approach of checking the consistency of updates to partial locations that may be auxiliary functions referring to parts of a complex object at flexible abstraction levels. Two kinds of partial updates were differentiated: shared updates and exclusive updates. This enabled us to develop a systematic approach for handling the compatibility issues of partial updates in general. In this approach, partial updates are processed in two stages: normalisation of shared updates and integration of exclusive updates. Our approach is related to applicative algebras [89]. However, we adopt - making it effective while discarding accidental cases - this to the context of complex-value databases which associate nested data structures with various type constructors.

7.2 Discussion

Let us first revisit the important differences between the postulates of database transformations, and Gurevich's sequential and parallel theses [27, 82].

Regarding states, we stayed with Gurevich's fundamental idea that states are first-order structures by generalising them with the notion of meta-finite states [74]. Meta-finite states are composed of a finite database part and a usually infinite algorithmic part with bridge functions linking them. Same as for the sequential ASM-thesis, closure under isomorphisms is requested for meta-finite states of a database transformation. By the incorporation of meta-finite states, we laid a solid foundation for further studying database transformations.

In order to capture different data models, we originally thought of manipulating the notion of state. In [154], the first attempt was the employment of higher-order structures to capture tree-structured databases such as object-oriented and XML-based databases. In this dissertation, we proposed that different data models should be captured by means of different background structures, which led us to formulate the background postulate. The results achieved so far on the backgrounds tailored for the relational, nested-relational, object-oriented and tree-based models show that shifting specific data model requirements to background structures is indeed effective.

The permission of non-determinism is significant for enhancing the expressive power of database transformations. On the other hand, every database transformation should also obey the genericity principle to respect the abstract nature of computations. This means that the degree of non-determinism allowed in database transformations needs to be restricted in some way. By the genericity postulate, we defined all the permissible successor states to a current state in terms of equivalence class of substructures. In doing so, the notion of semi-determinism defined for queries is generalised to generic updates, and thus to database transformations in general.

Though for database transformations the projection of a run onto the database part of states is most decisive, we did not build such a restriction into our model. This in turn implies that all sequential algorithms are also captured by DB-ASMs. Due to the permission of non-determinism that is restricted to choice among query results, we capture more than sequential algorithms. Even without the non-determinism this is still the case because of the more general bounded exploration witnesses and the exploitation of location operators, which permit a limited form of parallelism by aggregating multisets of values. Parallel algorithms, however, are not captured, as we only permit bounded parallelism as in the sequential ASM thesis.

The notion of database transformation developed in this paper is limited to sequential database transformations over centralised databases. That is, aspects of parallel and distributed databases have been neglected. We only considered parallelism for updates aggregated by means of location operators (i.e. we accumulate a multiset of updates on a location and then let them collapse to a single update). This form of parallelism is limited to the same computation on different data. Capturing parallelism and distribution as used in the architecture in [102] would require investigation of a more elaborate DB ASM thesis picking up ideas from the parallel ASM thesis [27].

The second-order syntax has been used in the logic for DB-ASMs, which permits the quantifications over a family of update sets and multisets yielded by a DB-ASM rule. Nevertheless, we showed that second-order expressions are reducible to first-order expressions because of the finiteness of updates in an update set or multiset, which is ensured by the finiteness of the database part in a state and the restrictions of all variables ranging only over in the database part. In doing so, the logic for DB-ASMs can indeed handle the bounded non-determinism. Moreover, we proved that the logic for DB-ASMs is still complete when the parameterised logic in the logic of meta-finite states is chosen to be the first-order logic.

Our approach for solving the problem of partial updates is generally applicable in complex-value databases. The normalisation of shared updates is purely based on algebraic properties of operators provided via the schema information, which can greatly improve the scalability of database applications. On the other hand, the integration of exclusive updates was studied in the presence of common type constructors occurring in complex-value databases. By regarding a data model in terms of arbitrary nesting of selected type constructors over base domains, the integration procedure may be tailored for many specific data models. Furthermore, the extension of locations to the use of auxiliary functions as partial locations not only allows to specify database transformations at more natural and flexible abstraction levels, but also facilitates more parallel data manipulations. It has been demonstrated that aggregate computing essentially leads to the presence of partial updates that may occur on the fly. This is one of promising applications for partial updates.

7.3 Future Work

In this dissertation, instead of requesting the existence of a finite set of ground terms that determines update sets (or families of update sets due to the assumed non-determinism), we widened this to a set of access terms, which in a sense captures associative access that is considered to be essential for databases. Nevertheless, there is still a big gap between access terms and similarly the access conditions in choice and forall rules in DB-ASMs on one side, and highly declarative query languages on the other side. The logical links would be of particular interest for queries, for which declarative approaches are preferred. Bringing these different aspects into alignment is a challenge for future research.

Moreover, we believe it is an interesting idea to consider "playing" with the notion of states, possibly even outside the context of databases. In particular, we are interested in states that can be recognised by certain types of automata. This may lead to establishing links between ASMs, database theory and particular structures, such as, automatic structures. For tree-based databases, we may think of structures that can be recognised by certain tree-automata or we could investigate automatic structures that are recognised by finite automata, etc. While these define further restrictions to the general computational model, they may provide interesting links to various logics, and also identify the challenge to integrate these automata into the formalism of DB-ASMs in order to capture exactly a particular class of database transformations.

Just like how ASMs were exploited to develop and verify Java and the JVM, another interesting line of research is to establish the links between DB-ASMs with tree-based backgrounds and other computation models for XML. Let us take XQuery for example. We may develop a low-level machine model for this industry standard language and then investigate the tool support for the verification of this class of computations (i.e., using a subclass of DB-ASMs to specify and verify XQuery). Furthermore, we may develop skeletons for initial specifications, standard refinement rules, etc. for XML machines so as to use XML machines as a specification tool allowing for XML database transformations specified at different abstraction levels.

With the incorporation of DB-ASMs in a theory of data-intensive web services, we may extend our results about database transformations into the area of service engineering, cloud computing, and in general service science. Using the methodology of ASMs, we may continue to characterise service mediators by using specific ASMs. Based on the characterisation of service mediators, further investigations can be conducted on the issues of analyzing the compatibility of services in service composition, developing dynamic and adaptive capabilities in service mediation, automating service composition and verifying the correctness, etc. There are a lot of challenging research problems in this area.

Bibliography

[1] Abiteboul, S., Hull, R., and Vianu, V. *Foundations of Databases*. Addison-Wesley, 1995.

[2] Abiteboul, S., and Kanellakis, P. C. Object identity as a query language primitive. In *Proceedings of the 1989 ACM SIGMOD international conference on Management of data* (New York, NY, USA, 1989), ACM Press, pp. 159–173.

[3] Abiteboul, S., and Kanellakis, P. C. Object identity as a query language primitive. *Journal of the ACM 45*, 5 (1998), 798–842.

[4] Abiteboul, S., Papadimitriou, C. H., and Vianu, V. The power of reflective relational machines. In *Logic in Computer Science* (1994), pp. 230–240.

[5] Abiteboul, S., Papadimitriou, C. H., and Vianu, V. Reflective relational machines. *Information and Computation 143*, 2 (1998), 110–136.

[6] Abiteboul, S., Quass, D., McHugh, J., Widom, J., and Wiener, J. L. The Lorel query language for semistructured data. *International Journal on Digital Libraries 1* (1997), 68–88.

[7] Abiteboul, S., Vardi, M. Y., and Vianu, V. Computing with infinitary logic. *Theoretical Computer Science 149*, 1 (1995), 101–128.

[8] Abiteboul, S., Vardi, M. Y., and Vianu, V. Fixpoint logics, relational machines, and computational complexity. *Journal of the ACM 44*, 1 (1997), 30–56.

[9] Abiteboul, S., and Vianu, V. Transaction languages for database update and specification. Tech. Rep. RR-0715, 1987.

[10] Abiteboul, S., and Vianu, V. Datalog extensions for database queries and updates. Tech. Rep. RR-0900, 1988.

[11] Abiteboul, S., and Vianu, V. Procedural languages for database queries and updates. *Journal of Computer and System Sciences 41*, 2 (1990), 181–229.

[12] Abiteboul, S., and Vianu, V. Datalog extensions for database queries and updates. *Journal of Computer and System Sciences 43*, 1 (1991), 62–124.

[13] Abiteboul, S., and Vianu, V. Generic computation and its complexity. In *Proceedings of the twenty-third annual ACM symposium on Theory of computing* (New York, NY, USA, 1991), ACM Press, pp. 209–219.

[14] Abiteboul, S., and Vianu, V. Computing with first-order logic. *Journal of Computer and System Sciences 50*, 2 (1995), 309–335.

[15] Aho, A. V., and Ullman, J. D. Universality of data retrieval languages. In *Proceedings of the 6th ACM SIGACT-SIGPLAN symposium on Principles of programming languages* (New York, NY, USA, 1979), ACM, pp. 110–119.

[16] Alon, N., Milo, T., Neven, F., Suciu, D., and Vianu, V. XML with data values: typechecking revisited. *Journal of Computer and System Sciences 66*, 4 (2003), 688–727.

[17] Andries, M., and Paredaens, J. On instance-completeness for database query languages involving object creation. *Journal of Computer and System Sciences 52*, 2 (1996), 357–373.

[18] Appendix, W. A., and Gurevich, Y. Interactive algorithms 2005. In *Interactive Computation: The New Paradigm* (2006), Springer.

[19] Bancilhon, F. On the completeness of query languages for relational data bases. In *Proceedings of the 7th International Symposium on Mathematical Foundations of Computer Science* (1978), pp. 112–123.

[20] Beeri, C., Milo, T., and Ta-Shma, P. On genericity and parametricity (extended abstract). In *Proceedings of the Fifteenth ACM SIGACT-SIGMOD-SIGART Symposium on Principles of Database Systems* (New York, NY, USA, 1996), ACM, pp. 104–116.

[21] Beeri, C., and Thalheim, B. Identification as a primitive of data models. In *Fundamentals of Information Systems*, T. Polle, T. Ripke, and K.-D. Schewe, Eds. Kluwer Academic Publishers, Boston Dordrecht London, 1999, pp. 19–36.

[22] BELLA, G., AND RICCOBENE, E. A realistic environment for crypto-protocol analyses by ASMs. In *Proceedings of the 5th International Workshop on Abstract State Machines* (1998), pp. 127–138.

[23] BENEDIKT, M., BONIFATI, A., FLESCA, S., AND VYAS, A. Adding updates to XQuery: Semantics, optimization, and static analysis. In *Proceedings of the Second International Workshop on XQuery Implementation, Experience and Perspectives* (2005).

[24] BERGLUND, A., BOAG, S., CHAMBERLIN, D., FERNÁNDEZ, M. F., KAY, M., ROBIE, J., AND SIMÉON, J. *XML Path Language (XPath) 2.0.* W3C Recommendation. September 2005. http://www.w3.org/TR/2005/WD-xpath20-20050915/.

[25] BIRON, P. V., AND MALHOTRA, A. *XML Schema Part 2: Datatypes*, second ed. W3C Recommendation. October 2004. http://www.w3.org/TR/xmlschema-2/.

[26] BLASS, A., AND GUREVICH, Y. Background, reserve, and Gandy machines. In *Proceedings of the 14th Annual Conference of the EACSL on Computer Science Logic* (London, UK, 2000), Springer-Verlag, pp. 1–17.

[27] BLASS, A., AND GUREVICH, Y. Abstract state machines capture parallel algorithms. *ACM Transactions on Computational Logic 4*, 4 (October 2003), 578–651.

[28] BLASS, A., AND GUREVICH, Y. Ordinary interactive small-step algorithms, I. *ACM Transactions on Computational Logic 7*, 2 (2006), 363–419.

[29] BLASS, A., AND GUREVICH, Y. Background of computation. *Bulletin of the European Association for Theoretical Computer Science (EATCS) 92* (June 2007).

[30] BLASS, A., AND GUREVICH, Y. Ordinary interactive small-step algorithms, II. *ACM Transactions on Computational Logic 8*, 3 (2007), 15.

[31] BLASS, A., AND GUREVICH, Y. Ordinary interactive small-step algorithms, III. *ACM Transactions on Computational Logic 8*, 3 (2007), 16.

[32] BLASS, A., AND GUREVICH, Y. Abstract state machines capture parallel algorithms: Correction and extension. *ACM Transactions on Computation Logic 9*, 3 (06 2008), 1–32.

[33] Blass, A., and Gurevich, Y. Persistent queries. *The Computing Research Repository abs/0811.0819* (2008).

[34] Blass, A., Gurevich, Y., Rosenzweig, D., and Rossman, B. Interactive small-step algorithms I: Axiomatization. *Logical Methods in Computer Science 3*, 4:3 (2007), 1–29.

[35] Blass, A., Gurevich, Y., Rosenzweig, D., and Rossman, B. Interactive small-step algorithms II: Abstract state machines and the characterization theorem. *Logical Methods in Computer Science 3*, 4:4 (2007), 1–35.

[36] Blass, A., Gurevich, Y., and Shelah, S. Choiceless polynomial time. *Annals of Pure and Applied Logic 100* (1999), 141–187.

[37] Blass, A., Gurevich, Y., and Shelah, S. On polynomial time computation over unordered structures. *The Journal of Symbolic Logic 67*, 3 (September 2002), 1093–1125.

[38] Blass, A., Gurevich, Y., and Van den Bussche, J. Abstract state machines and computationally complete query languages. *Information and Computation 174*, 1 (April 2002), 20–36.

[39] Bloem, R., and Engelfriet, J. Characterization of properties and relations defined in monadic second order logic on the nodes of trees. Tech. rep., 1997.

[40] Boag, S., Chamberlin, D., Fernéndez, M. F., Florescu, D., Robie, J., and Siméon, J. *XQuery 1.0: An XML Query Language.* W3C Recommendation. January 2007. http://www.w3.org/TR/xquery.

[41] Bojanczyk, M., and Walukiewicz, I. Forest algebras. In *Automata and Logic: History and Perspectives* (2007), Amsterdam University Press, pp. 107–132.

[42] Bonifati, A., and Ceri, S. Comparative analysis of five XML query languages. *ACM SIGMOD Record 29*, 1 (2000), 68–79.

[43] Bonner, A. J., and Kifer, M. Transaction logic programming. In *Proceedings of the tenth international conference on logic programming on Logic programming* (Cambridge, MA, USA, 1993), MIT Press, pp. 257–279.

[44] BONNER, A. J., AND KIFER, M. An overview of transaction logic. *Theoretical Computer Science 133*, 2 (1994), 205–265.

[45] BONNER, A. J., AND KIFER, M. The state of change: A survey. In *International Seminar on Logic Databases and the Meaning of Change, Transactions and Change in Logic Databases* (London, UK, 1998), Springer-Verlag, pp. 1–36.

[46] BONNER, A. J., KIFER, M., AND CONSENS, M. P. Database programming in transaction logic. In *Proceedings of the Fourth International Workshop on Database Programming Languages - Object Models and Languages* (London, UK, 1994), Springer-Verlag, pp. 309–337.

[47] BÖRGER, E., AND STÄRK, R. F. *Abstract State Machines: A Method for High-Level System Design and Analysis.* Springer-Verlag New York, Inc., Secaucus, NJ, USA, 2003.

[48] BÜCHI, J. R. Weak second-order arithmetic and finite automata. *Zeitschrift für Mathematische Logik und Grundlagen der Mathematik 6*, 1-6 (1960), 66–92.

[49] CAI, J.-Y., FÜRER, M., AND IMMERMAN, N. An optimal lower bound on the number of variables for graph identification. In *Proceedings of the 30th Annual Symposium on Foundations of Computer Science* (Washington, DC, USA, 1989), IEEE Computer Society, pp. 612–617.

[50] CHAMBERLIN, D., CAREY, M., FLORESCU, D., KOSSMANN, D., AND ROBIE, J. XQueryP: Programming with XQuery. In *Proceedings of the third International Workshop on XQuery Implementation, Experience, and Perspectives* (2006).

[51] CHAMBERLIN, D., FLORESCU, D., MELTON, J., ROBIE, J., AND SIMÉON, J. *XQuery Update Facility 1.0.* W3C Recommendation. August 2008. http://www.w3.org/TR/xquery-update-10.

[52] CHANDRA, A. K., AND HAREL, D. Computable queries for relational data bases. *Journal of Computer and System Sciences 21*, 2 (1980), 156–178.

[53] CLARK, J. *XSL Transformations (XSLT).* W3C Recommendation. November 1999. http://www.w3.org/TR/xslt.

[54] Clark, J., and Makoto, M. *RELAX NG Specification.* OASIS, December 2001. http://www.oasis-open.org/committees/relax-ng/spec-20011203.html.

[55] Clark, J., and Makoto, M. *RELAX NG Tutorial.* OASIS, December 2001. http://relaxng.org/tutorial-20011203.html.

[56] Codd, E. F. Derivability, redundancy and consistency of relations stored in large data banks. *IBM Research Report, San Jose, California RJ599* (1969).

[57] Codd, E. F. A relational model of data for large shared data banks. *Communications of the ACM 13*, 6 (1970), 377–387.

[58] Codd, E. F. Further normalization of the data base relational model. *IBM Research Report, San Jose, California RJ909* (1971).

[59] Codd, E. F. Relational completeness of data base sublanguages. *Database Systems: 65-98, Prentice Hall and IBM Research Report RJ 987, San Jose, California* (1972).

[60] Cohen, S. User-defined aggregate functions: bridging theory and practice. In *Proceedings of the 2006 ACM SIGMOD international conference on Management of data* (New York, NY, USA, 2006), ACM Press, pp. 49–60.

[61] Comon, H., Dauchet, M., Gilleron, R., Löding, C., Jacquemard, F., Lugiez, D., Tison, S., and Tommasi, M. Tree automata techniques and applications, 2007. http://www.grappa.univ-lille3.fr/tata.

[62] Dahlhaus, E., and Makowsky, J. A. Query languages for hierarchic databases. *Information and Computation 101*, 1 (1992), 1–32.

[63] Denninghoff, K., and Vianu, V. Database method schemas and object creation. In *Proceedings of the twelfth ACM SIGACT-SIGMOD-SIGART symposium on Principles of database systems* (New York, NY, USA, 1993), ACM, pp. 265–275.

[64] Deutsch, A., Fernández, M. F., Florescu, D., Levy, A. Y., and Suciu, D. A query language for XML. *Computer Networks 31*, 11-16 (1999), 1155–1169.

[65] Doner, J. Tree acceptors and some of their applications. *Journal of Computer and System Sciences 4*, 5 (1970), 406–451.

[66] EBBINGHAUS, H.-D., AND FLUM, J. *Finite Model Theory*, 2nd ed. Springer-Verlag, 1999.

[67] ENDERTON, H. *A mathematical introduction to logic.* Academic Press, Inc., New York, USA, 1972.

[68] FAGIN, R. Finite-model theory - a personal perspective. In *Proceedings of the third international conference on database theory on Database theory* (New York, NY, USA, 1990), Springer-Verlag, pp. 3–24.

[69] FENSEL, D., AND GROENBOOM, R. MLPM: Defining a semantics and axiomatization for specifying the reasoning process of knowledge-based systems. In *Proceedings of the 12th European Conference on Artificial Intelligence (ECAI-96)* (Budapest, Hungary, 1996).

[70] GHELLI, G., RE, C., AND SIMÉON, J. XQuery!: An XML query language with side effects. In *EDBT Workshops* (2006), pp. 178–191.

[71] GLÄSSER, U., AND VEANES, M. Universal plug and play machine models. In *Proceedings of the IFIP 17th World Computer Congress - TC10 Stream on Distributed and Parallel Embedded Systems* (Deventer, The Netherlands, The Netherlands, 2002), Kluwer, B.V., pp. 21–30.

[72] GLAUSCH, A., AND REISIG, W. An ASM-Characterization of a Class of Distributed Algorithms. In *Proceedings of the Dagstuhl Seminar on Rigorous Methods for Software Construction and Analysis* (2007), Festschrift volume of Lecture Notes in Computer Science, Springer.

[73] GRÄDEL, E. Finite model theory and descriptive complexity. In *Finite Model Theory and Its Applications*. Springer-Verlag, 2003.

[74] GRÄDEL, E., AND GUREVICH, Y. Metafinite model theory. *Information and Computation 140*, 1 (1998), 26–81.

[75] GRÄDEL, E., AND OTTO, M. Inductive definability with counting on finite structures. In *Selected Papers from the Workshop on Computer Science Logic* (London, UK, 1993), Springer-Verlag, pp. 231–247.

[76] Groenboom, R., and Renardel de Lavalette, G. Reasoning about dynamic features in specification languages - a modal view on creation and modification. In *Proceedings of the International Workshop on Semantics of Specification Languages (SoSL)* (London, UK, 1994), Springer-Verlag, pp. 340–355.

[77] Groenboom, R., and Renardel de Lavalette, G. A formalization of evolving algebras. In *Proceedings of Accolade95* (1995), Dutch Research School in Logic.

[78] Gurevich, Y. A new thesis (abstracts). *American Mathematical Society 6*, 4 (August 1985), 317.

[79] Gurevich, Y. Logic and the challenge of computer science. *Current Trends in Theoretical Computer Science* (1988), 1–57.

[80] Gurevich, Y. Evolving algebras 1993: Lipari guide. In *Specification and Validation Methods* (1993), Oxford University Press, pp. 9–36.

[81] Gurevich, Y. May 1997 draft of the ASM guide. Tech. Report CSE-TR-336-97, EECS Department, University of Michigan, 1997.

[82] Gurevich, Y. Sequential abstract state machines capture sequential algorithms. *ACM Transactions on Computational Logic 1*, 1 (July 2000), 77–111.

[83] Gurevich, Y. Intra-step interaction. In *Abstract State Machines* (2004), pp. 1–5.

[84] Gurevich, Y., and Rosenzweig, D. Partially ordered runs: A case study. In *Proceedings of the International Workshop on Abstract State Machines, Theory and Applications* (London, UK, 2000), Springer-Verlag, pp. 131–150.

[85] Gurevich, Y., Rossman, B., and Schulte, W. Semantic essence of AsmL. *Theoretical Computer Science 343*, 3 (2005), 370–412.

[86] Gurevich, Y., Schulte, W., and Veanes, M. Rich sequential-time ASMs. In *Formal Methods and Tools for Computer Science* (Canary Islands, Spain, 2001), Universidad de Las Palmas de Gran Canaria, pp. 291–293.

[87] Gurevich, Y., and Tillmann, N. Partial updates: Exploration. *Journal of Universal Computer Science 7*, 11 (November 2001), 917–951.

[88] Gurevich, Y., and Tillmann, N. Partial updates exploration II. In *Abstract State Machines* (2003).

[89] Gurevich, Y., and Tillmann, N. Partial updates. *Theoretical Computer Science 336*, 2-3 (2005), 311–342.

[90] Gurevich, Y., and Yavorskaya, T. On bounded exploration and bounded nondeterminism. Tech. Rep. MSR-TR-2006-07, Microsoft Research, January 2006.

[91] Gyssens, M., Paredaens, J., Van den Bussche, J., and Van Gucht, D. A graph-oriented object database model. *IEEE Transactions on Knowledge and Data Engineering 6*, 4 (1994), 572–586.

[92] Gyssens, M., Van den Bussche, J., and Van Gucht, D. Expressiveness of efficient semi-deterministic choice constructs. In *Proceedings of the 21st International Colloquium on Automata, Languages and Programming* (London, UK, 1994), Springer-Verlag, pp. 106–117.

[93] Hella, L., Libkin, L., Nurmonen, J., and Wong, L. Logics with aggregate operators. *Journal of the ACM 48*, 4 (2001), 880–907.

[94] Huggins, J. Abstract state machines web page. http://www.eecs.umich.edu/gasm.

[95] Hughes, G., and Cresswell, M. *A new introduction to modal logic.* Burns & Oates, 1996.

[96] Hull, R., and Su, J. Untyped sets, invention, and computable queries. In *Proceedings of the eighth ACM SIGACT-SIGMOD-SIGART symposium on Principles of database systems* (New York, NY, USA, 1989), ACM, pp. 347–359.

[97] Hull, R., and Su, J. Algebraic and calculus query languages for recursively typed complex objects. *Journal of Computer and System Sciences 47*, 1 (1993), 121–156.

[98] Hull, R., and Yap, C. K. The format model: a theory of database organization. In *Proceedings of the 1st ACM SIGACT-SIGMOD symposium on Principles of database systems* (New York, NY, USA, 1982), ACM, pp. 205–211.

[99] Immerman, N. Relational queries computable in polynomial time. *Information and Control 68*, 1-3 (1986), 86–104.

[100] IMMERMAN, N. Expressibility as a complexity measure: Results and directions. In *Proceedings of Second Conference on Structure in Complexity Theory* (1987), pp. 194–202.

[101] KIRCHBERG, M. Using abstract state machines to model ARIES-based transaction processing. *Journal of Universal Computer Science (J.UCS) 15*, 1 (2009), 157–194.

[102] KIRCHBERG, M., SCHEWE, K.-D., TRETIAKOV, A., AND WANG, R. A multi-level architecture for distributed object bases. *Data & Knowledge Engineering 60*, 1 (2007), 150–184.

[103] KOLMOGOROV, A. N., AND USPENSKII, V. A. On the definition of an algorithm. In *Uspekhi Mat. Nauk* (Russian, 1958), vol. 13:4, pp. 3–28.

[104] LEE, D., AND CHU, W. W. Comparative analysis of six XML schema languages. *ACM SIGMOD Record 29*, 3 (2000), 76–87.

[105] LEVENE, M. *The Nested Universal Relation Database Model.* Springer, 1992.

[106] LIBKIN, L. Logics with counting and local properties. *ACM Transactions on Computational Logic 1*, 1 (2000), 33–59.

[107] MA, H., SCHEWE, K.-D., THALHEIM, B., AND WANG, Q. Abstract state services. In *Proceedings of the ER 2008 Workshops on Advances in Conceptual Modeling* (Berlin, Heidelberg, 2008), Springer-Verlag, pp. 406–415.

[108] MA, H., SCHEWE, K.-D., THALHEIM, B., AND WANG, Q. A theory of data-intensive software services. *Service Oriented Computing and Applications 3*, 4 (2009), 263–283.

[109] MA, H., SCHEWE, K.-D., AND WANG, Q. An abstract model for service provision, search and composition. In *Proceedings of 2009 IEEE Asia-Pacific Services Computing Conference* (Singapore, 2009).

[110] MAKINOUCHI, A. A consideration of normal form of not-necessarily-normalized relations in the relational database model. In *Proceedings of International Conference on Very Large Data Bases* (1977), pp. 447–453.

[111] MURATA, M., LEE, D., MANI, M., AND KAWAGUCHI, K. Taxonomy of XML schema languages using formal language theory. *ACM Transactions on Internet Technology 5*, 4 (2005), 660–704.

[112] NANCHEN, S. *Verifying Abstract State Machines.* PhD thesis, ETH Zurich, Switzerland, 2007.

[113] NÉMETH, Z., AND SUNDERAM, V. A Formal Framework for Defining Grid Systems. In *Proceedings of the Second IEEE/ACM International Symposium on Cluster Computing and the Grid* (May 2002), IEEE Computer Society Press, pp. 202–211.

[114] OTTO, M. The expressive power of fixed-point logic with counting. *Journal of Symbolic Logic 61* (1996), 147–176.

[115] OTTO, M. *Bounded variable logics and counting – A study in finite models*, vol. 9. Springer-Verlag, 1997.

[116] PAPAKONSTANTINOU, Y., AND VIANU, V. DTD Inference for Views of XML Data. In *Proceedings of the Nineteenth ACM SIGMOD-SIGACT-SIGART Symposium on Principles of Database Systems* (Dallas, Texas, 2000), pp. 35–46.

[117] PAREDAENS, J. On the expressive power of the relational algebra. *Information Processing Letters 7*, 2 (1978), 107–111.

[118] RAMAKRISHNAN, R., ROSS, K. A., SRIVASTAVA, D., AND SUDARSHAN, S. Efficient incremental evaluation of queries with aggregation. In *Proceedings of the 1994 International Symposium on Logic programming* (Cambridge, MA, USA, 1994), MIT Press, pp. 204–218.

[119] REISIG, W. On Gurevich's theorem on sequential algorithms. *Acta Informatica 39*, 5 (2003), 273–305.

[120] RENARDEL DE LAVALETTE, G. A logic of modification and creation. In *Logical Perspectives on Language and Information. CSLI publications* (2001).

[121] ROBIE, J., LAPP, J., AND SCHACH, D. XML query language (XQL). In *Proceedings of the Query Languages workshop* (Cambridge, Mass, Dec. 1998).

[122] ROSENZWEIG, D., RUNJE, D., AND SLANI, N. Privacy, abstract encryption and protocols: An ASM model - part I. In *Abstract State Machines* (2003), Springer Berlin / Heidelberg, pp. 372–390.

[123] ROTH, M. A., KORTH, H. F., AND SILBERSCHATZ, A. Theory of non-first-normal form relational databases. Tech. Rep. TR-84-36, University of Texas, Austin, Texas, USA, 1986.

[124] SCHEWE, K.-D., AND THALHEIM, B. Fundamental concepts of object oriented databases. *Acta Cybernetica 11*, 1-2 (1993), 49–84.

[125] SCHEWE, K.-D., THALHEIM, B., AND WANG, Q. Validation of streaming XML documents with abstract state machines. In *Proceedings of the 10th International Conference on Information Integration and Web-based Applications and Services* (2008), pp. 147–153.

[126] SCHEWE, K.-D., THALHEIM, B., AND WANG, Q. Updates, schema updates and validation of XML documents - using abstract state machines with automata-defined states. *Journal of Universal Computer Science 15*, 10 (2009), 2028–2057.

[127] SCHÖNEGGE, A. Extending Dynamic Logic for Reasoning about Evolving Algebras. Technical Report 49/95, Universität Karlsruhe, Fakultät für Informatik, 1995.

[128] SCHÖNHAGE, A. Storage modification machines. In *Proceedings of the 4th GI-Conference on Theoretical Computer Science* (London, UK, 1979), Springer-Verlag, pp. 36–37.

[129] SEGOUFIN, L., AND VIANU, V. Validating streaming XML documents. In *Proceedings of the 21st ACM SIGACT-SIGMOD-SIGART Symposium on Principles of Database Systems*. ACM, 2002, pp. 53–64.

[130] SPRUIT, P. *Logics of Database Updates*. PhD thesis, Faculty of Mathematics and Computer Science, Vrije Universiteit, Amsterdam, 1994.

[131] SPRUIT, P., WIERINGA, R., AND MEIJER, J.-J. Axiomatization, declarative semantics and operational semantics of passive and active updates in logic databases. *Journal of Logic and Computation 5* (1995), 27–50.

[132] Spruit, P., Wieringa, R., and Meyer, J.-J. Dynamic database logic: The first-order case. In *Modelling Database Dynamics* (1993), U. Lipeck and B. Thalheim, Eds., Springer, pp. 103–120.

[133] Spruit, P., Wieringa, R., and Meyer, J.-J. Regular database update logics. *Theoretical Computer Science 254*, 1-2 (2001), 591–661.

[134] Stärk, R., and Nanchen, S. A logic for abstract state machine. *Jurnal of Universal Computer Science 7*, 11 (2001).

[135] Stärk, R., Schmid, J., and Börger, E. *Java and the Java Virtual Machine: Definition, Verification, Validation.* Springer-Verlag, 2001.

[136] Suciu, D. Domain-independent queries on databases with external functions. *Theoretical Computer Science 190*, 2 (1998), 279–315.

[137] Sur, G., Hammer, J., and Siméon, J. An XQuery-based language for processing updates in XML. *Proceedings of Programming Language Technologies for XML* (2004).

[138] ten Cate, B. The expressivity of XPath with transitive closure. In *Proceedings of the twenty-fifth ACM SIGMOD-SIGACT-SIGART symposium on Principles of database systems* (New York, NY, USA, 2006), ACM, pp. 328–337.

[139] ten Cate, B., and Segoufin, L. XPath, transitive closure logic, and nested tree walking automata. In *Proceedings of the twenty-seventh ACM SIGMOD-SIGACT-SIGART symposium on Principles of database systems* (New York, NY, USA, 2008), ACM, pp. 251–260.

[140] Thatcher, J. W., and Wright, J. B. Generalized finite automata theory with an application to a decision problem of second-order logic. *Mathematical Systems Theory 2*, 1 (1968), 57–81.

[141] Thomas, S. J., and Fischer, P. C. Nested relational structures. *Advances in Computing Research 3* (1986), 269–307.

[142] Thompson, H. S., Beech, D., Maloney, M., and Mendelsohn, N. *XML Schema Part 1: Structures*, second ed. W3C Recommendation. October 2004. http://www.w3.org/TR/xmlschema-1/.

[143] TURING, A. M. On computable numbers, with an application to the Entscheidungsproblem. *Proceedings of The London Mathematical Society 2*, 42 (1936), 230–265.

[144] TURULL TORRES, J. M. On the expressibility and the computability of untyped queries. *Annals of Pure and Applied Logic 108*, 1-3 (2001), 345–371.

[145] TURULL TORRES, J. M. Relational databases and homogeneity in logics with counting. *Acta Cybernetica 17*, 3 (2006), 485–511.

[146] VAN DEN BUSSCHE, J. *Formal Aspects of Object Identity in Database Manipulation.* PhD thesis, University of Antwerp, 1993.

[147] VAN DEN BUSSCHE, J. Tree-structured object creation in database transformations. *the Liber Amicorum for Jan Paredaens's 60th Birthday* (21 September 2007).

[148] VAN DEN BUSSCHE, J., AND VAN GUCHT, D. Semi-determinism (extended abstract). In *Proceedings of the eleventh ACM SIGACT-SIGMOD-SIGART symposium on Principles of database systems* (New York, NY, USA, 1992), ACM Press, pp. 191–201.

[149] VAN DEN BUSSCHE, J., AND VAN GUCHT, D. Non-deterministic aspects of object-creating database transformations. In *Selected Papers from the Fourth International Workshop on Foundations of Models and Languages for Data and Objects* (London, UK, 1993), Springer-Verlag, pp. 3–16.

[150] VAN DEN BUSSCHE, J., VAN GUCHT, D., ANDRIES, M., AND GYSSENS, M. On the completeness of object-creating database transformation languages. *Journal of the ACM 44*, 2 (1997), 272–319.

[151] VAN ECK, P., ENGELFRIET, J., FENSEL, D., VAN HARMELEN, F., VENEMA, Y., AND WILLEMS, M. A survey of languages for specifying dynamics: A knowledge engineering perspective. *IEEE Transactions on Knowledge and Data Engineering 13*, 3 (2001), 462–496.

[152] VARDI, M. Y. The complexity of relational query languages (extended abstract). In *Proceedings of the fourteenth annual ACM symposium on Theory of computing* (New York, NY, USA, 1982), ACM, pp. 137–146.

[153] WANG, Q., AND FERRAROTTI, F. A. XML machines. In *Lecture Notes in Computer Science* (2009), vol. 5833, Springer, pp. 95–104.

[154] WANG, Q., AND SCHEWE, K.-D. Axiomatization of database transformations. In *Proceedings of the 14th International ASM Workshop* (2007).

[155] WANG, Q., AND SCHEWE, K.-D. Towards a logic for abstract metafinite state machines. In *Lecture Notes in Computer Science* (2008), S. Hartmann and G. Kern-Isberner, Eds., vol. 4932, Springer, pp. 365–380.

[156] WANG, Q., SCHEWE, K.-D., AND THALHEIM, B. XML database transformations with tree updates. In *Lecture Notes in Computer Science* (2008), vol. 5238, Springer, p. 342.

[157] WILKE, T. An algebraic characterization of frontier testable tree languages. *Theoretical Computer Science 154*, 1 (1996), 85–106.

[158] YAN, W. P., AND LARSON, P.-A. Eager aggregation and lazy aggregation. In *Proceedings of the 21th International Conference on Very Large Data Bases* (San Francisco, CA, USA, 1995), Morgan Kaufmann Publishers Inc., pp. 345–357.

List of Figures

List of Symbols

S	a state (or a structure)	Definition 3.1.4
$Aut(S)$	automorphism group of S	page 22
$Adom(S)$	active domain of S	page 22
D	a domain	page 35
ς	isomorphism between two structures	Definition 3.1.5
Z	a set of constants	page 40
a, b, c	constants	page 40
$t_1, t_2, \ldots$	terms	page 40
$val_S(t)$	interpretation of ground t in S	page 40
T	a database transformation	Postulate 3.1.1
$\mathcal{S}_T$	non-empty set of states of T	Postulate 3.1.1
$\mathcal{I}_T$	set of initial states of T	Postulate 3.1.1
$\mathcal{F}_T$	set of final states of T	Postulate 3.1.1
δ_T	one-step transition relation of T	Postulate 3.1.1
Υ	a state signature	Definition 3.1.4
Υ_{db}, Υ_a	database and algorithmic parts of Υ	page 41
B	base set of a structure	Definition 3.1.4
B_{db}^{ext}, B_a	database and algorithmic parts of B	page 41
B_{db}	subset of B_{db}^{ext}	page 41
$o_1, o_2, \ldots$	object or node identifiers	pages 44, 93
$\mathcal{D}$	set of domains	page 45
U	a universe	page 45
$\lfloor\,\rfloor$	a constructor symbol	page 45
Υ_K	a background signature	page 45
K	a background class	Definition 3.1.7
$HF(D)$	set of hereditarily finite sets over D	page 46

$\tau_1, \tau_2, ...$	types or type expressions	page 46
$[\![\tau]\!]$	interpretation of τ	page 46
$\{\cdot\}$, $(\cdot)$, $[\cdot]$, $\{\!\!\{\cdot\}\!\!\}$	type constructors for set, tuple, list and multiset	pages 44 - 48
$\sqcup$, $\sqcap$	type constructors for union and intersection	page 47
α, β	function mappings	page 47
$\uplus$	binary multiset union	page 47
$\biguplus x$	general multiset union	page 47
$AsSet$	unary multiset operation returning a set	page 47
$\mathbf{I}x$	the unique element in x	page 47
$\{\!\!\{\ \}\!\!\}$	an empty multiset	page 48
$\{\!\!\{x\}\!\!\}$	a singleton multiset	page 48
$\bot$	undefinedness value	page 48
$\ell_1, \ell_2, ...$	locations	Definition 3.1.8
(ℓ, b)	an update	Definition 3.1.8
Δ	an update set	Definition 3.1.8
$\ddot{\Delta}$	an update multiset	Definition 3.1.8
$S + \Delta$	applying Δ to S	page 49
$\Delta(T, S)$	set of update sets yielded by T on S	page 49
$\Delta(T, S, S')$	a minimal update set yielded by T from S to S'	page 49
Loc_Δ	set of locations in Δ on which S and S' differ ($Loc_\Delta = \{\ell \mid val_S(\ell) \neq val_{S'}(\ell)\}$)	page 49
$\rho_1, \rho_2, ...$	location operators	Definition 3.1.9
θ	a location function	page 51
$\mathcal{M}(D)$	set of all non-empty multisets over D	page 51
φ, ψ	formulae	page 53
ζ	a variable assignment	page 55
$val_{S,\zeta}(t)$	interpretation of t in S under ζ	page 55
(t_β, t_α), t_α	access terms	Definition 3.1.12
$\mathcal{T}_{wit}$	a bounded exploration witness	page 56
$S' \preceq S$	S' is a substructure of S	Definition 3.1.13
$S' \equiv S$	S' and S are equivalent	page 57
r	a rule	page 60
$\Delta(r, S)$	set of update sets yielded by r on S	page 60
$\Delta(r, S, \zeta)$	set of update sets yielded by r on S under ζ	page 61
$\ddot{\Delta}(r, S)$	set of update multisets yielded by r on S	page 60
$\ddot{\Delta}(r, S, \zeta)$	set of update multisets yielded by r on S under ζ	page 61

$fr(r)$	set of free variables in r	page 61
$var(t)$	set of variables in t	page 61
$\mathcal{R}$	set of rules	page 61
$\mathcal{B}$	set of tuples of values	page 62
$\mathfrak{M}$	set of function mappings	page 62
$\Delta_1 \oslash \Delta_2$	sequential composition of Δ_1 and Δ_2	page 63
$\ddot{\Delta}_1 \oslash \ddot{\Delta}_2$	sequential composition of $\ddot{\Delta}_1$ and $\ddot{\Delta}_2$	page 63
M	a DB-ASM	Definition 3.2.2
$\mathcal{S}_M$	set of states of M	page 65
$\mathcal{I}_M$	set of initial states of M	page 65
$\mathcal{F}_M$	set of final states of M	page 65
r_M	a closed rule of M	page 65
δ_M	a binary relation of M	page 65
$t^{'}[t/x]$, $\varphi[t/x]$	substitution of term t for variable x in $t^{'}$ or φ	Definition 3.2.3
CT	set of critical terms	page 68
c_{ct}	a critical term	Definition 3.3.1
$\bar{C}_S$	set of critical elements of S	Definition 3.3.2
E_S	equivalence class on CT of S	Definition 3.3.3
u	an update	page 70
$\mathcal{L}$	set of locations	page 70
$\mathcal{L}_{pre}$, $\mathcal{L}_{post}$	two sets of relational locations corresponding to $\mathcal{L}$	page 70
E_{pre}	equivalence class of substructures defined by $\mathcal{L}_{pre}$	page 71
E_{post}	equivalence class of substructures defined by $\mathcal{L}_{post}$	page 71
$\overline{x}_1, \overline{x}_2, \ldots$	tuples of variables	page 72
R	a (nested) relation schema	page 81
$attr(R)$	set of attribute names of R	page 81
$I(R)$	a (nested) relation over R	page 81
$\Re$	set of (nested) relation schemata	page 81
$I(\Re)$	a (nested) relational database instance over $\Re$	page 81
dom	a domain assignment	page 81
A	an attribute name	page 81
$\mathcal{P}(D)$	the power set of D	page 81
$\mathcal{A}$	set of attribute names	page 83
Υ_K^{rel}	a relational background signature	Definition 4.1.1
K^{rel}	a relational background class	Definition 4.1.2
$\mathfrak{T}$	set of types	page 84
typ	a type assignment	page 84
$\tilde{\Gamma}^{rel}$	set of relational type schemes	page 85

$atom(t)$	set of all atoms in t	page 85
$term(K)$	set of all terms in K	page 86
$\widehat{ID}$	set of identities	page 85
$\sim^t$	t-equivalence of terms	page 86
$\sim^i$	i-equivalence of terms	page 86
σ	selection operator	page 86
π	projection operator	page 87
$\bowtie$	natural join operator	page 87
ϱ	renaming operator	page 87
$p_1, p_2, \ldots$	programs	page 90
γ	an unranked tree	Definition 4.2.1
$\mathcal{O}_\gamma$	tree domain of γ	page 93
$\prec_c$	child relation	page 93
$\prec_s$	sibling relation	page 93
Σ	a finite, non-empty set of labels	page 93
t_γ	an XML tree	Definition 4.2.2
ω	label function of an XML tree	page 93
υ	value function of an XML tree	page 93
$\mathcal{T}(\Sigma)$	set of all XML trees over Σ	page 94
$\widehat{o}$	a subtree of $t_\gamma = (\gamma, \omega, \upsilon)$ rooted at $o \in \mathcal{O}_\gamma$	page 93
$root(t_\gamma)$	root node of t_γ.	page 93
ε	an empty hedge	page 94
t_c	an XML context	Definition 4.2.3
$\mathcal{T}(\Sigma \cup \{\xi\})$	set of all XML contexts over Σ	page 94
ξ	a trivial context	page 94
$\prec_c^*$	transitive closure of $\prec_c$	page 95
$\#t$	the sort of t	page 98
L, **H**, **C**	sorts for labels, for hedges and for contexts	page 97
$f_\iota, f_\nu, f_\varsigma, f_\varrho$	function symbols in XML tree algebra	page 97
$f_\kappa, f_\eta, f_\vartheta$	function symbols in XML tree algebra	page 97
$\mathcal{T}_L$	set of label terms	page 98
$\mathcal{T}_H$	set of hedge terms	page 98
$\mathcal{T}_C$	set of context terms	page 98
$\mathcal{T}_H^s$	set of tree terms	page 98
$\mathcal{T}$	set of terms	page 98

$\mathcal{X}_{FO}$	set of first-order variables	page 103
$\mathcal{X}_{SO}$	set of second-order variables	page 103
$x_1, x_2, ...$	first-order variables	page 103
$X_1, X_2, ...$	second-order variables	page 103
$[\![\varphi]\!]_{S,\zeta}$	interpretation of φ in S under ζ	page 104
Υ_K^{tree}	a tree-based background signature	Definition 4.4.1
$HFT(\mathcal{O})$	set of all finite trees over $\mathcal{O}$	Definition 4.4.2
$HFT(\mathcal{O}, \xi)$	set of all contexts over $\mathcal{O}$	Definition 4.4.3
K^{tree}	a tree background class	Definition 4.4.4
$reg(\Sigma)$	set of regular languages over Σ	page 108
G	a DTD or an EDTD	page 110
$sat(G)$	set of XML trees satisfying G	page 111
typ_n	a type name assignment	page 111
typ_e	a type expression assignment	page 111
$\hat{\Gamma}^{tree}$	a set of tree type schemes	page 112
$\mathfrak{D}$	an XML database schema	page 112
$I(\mathfrak{D})$	an XML database instance over $\mathfrak{D}$	page 112
$\mathcal{R}_{MSO}$	set of MSO-rules	page 114
$\mathcal{T}_{db}$	set of database terms	page 122
$\mathcal{T}_a$	set of algorithmic terms	page 122
P	a predicate symbol	page 123
$\mathbb{F}_{dyn}$	set of dynamic function symbols in a signature	page 128
$(U_{\mathfrak{K}}, R_{\mathfrak{K}})$	a Kripke frame	Definition 5.3.1
$\Psi \models_M \varphi$	φ is implied by Ψ with respect to M	Definition 5.3.2
$\Psi \vdash_M \varphi$	φ is derived from Ψ with respect to M	Definition 5.3.3
$r_1 \simeq r_2$	r_1 and r_2 are equivalent	Definition 5.3.4
$\lvert\varphi\rvert$ $(\lvert r\rvert)$	the rank of φ (or r)	page 151
$\mathcal{X}_H$	a set of Henkin constants	page 152
$\sim_\Psi$	an equivalence relation on variables w.r.t. Ψ	page 155
$\lceil x \rceil_\Psi$	equivalence class of x	page 155
$\lceil \sim_\Psi \rceil$	set of all equivalence classes w.r.t. $\sim_\Psi$ in a state	page 155
$\widetilde{\zeta}$	a variable assignment assigning $\lceil x \rceil_\Psi$ to x	page 156

$\mathcal{N}(D)$	set of all finite lists over D	page 170
$\mathfrak{A}$	an applicative algebra	Definition 6.2.1
λ	a trivial element or particle	page 173
Ω	parallel composition operation	page 173
s	a sequence	page 174
$\sharp s$	the length of s	page 175
$Loc(\mathcal{M})$	set of locations occurring in $\mathcal{M}$	page 175
$\mathcal{M}_\ell$	$\mathcal{M}_\ell = \{\!\{(f,\ell) \mid (f,\ell) \in \mathcal{M}\}\!\}$	page 175
$\sqcup_\ell$, $\sqcap_\ell$	two binary operations over a set of locations	page 179
$\ell_2 \sqsubseteq \ell_1$	ℓ_1 subsumes ℓ_2	Definition 6.3.2
$\ell_2 \trianglelefteq \ell_1$	ℓ_1 depends on ℓ_2	Definition 6.3.4
$\tau(\ell)$	type of ℓ	page 182
μ	a binary operator in share update (ℓ, b, μ)	Definition 6.3.5
$Loc(\ddot{\Delta})$	set of locations occurring in $\ddot{\Delta}$	page 185
$Opt(\ddot{\Delta})$	set of operators occurring in $\ddot{\Delta}$	page 185
$\ddot{\Delta}_\ell$	submultiset of $\ddot{\Delta}$ containing all shared updates that have the location ℓ	page 185
$\mu_1 \preceq \mu_2$	μ_1 is compatible to μ_2	Definition 6.4.3
$norm(\ddot{\Delta})$	update set after the normalisation of $\ddot{\Delta}$	page 189
Δ_λ	a trivial update set	page 189
$SubL(\ell)$	set of all the sublocations of ℓ	page 194
$height(\Delta_{ij})$	height of a cluster Δ_{ij}	page 196
$clus(\Delta)$	set of clusters corresponding to Δ	page 200